AusTrALiA 2030!

WHERE the bloody hell are we?

Life is anything but usual, and uncertainty
has become the new normal.

Do we take 'the road less travelled',
or 'the road not taken' and what is
on the other side anyway?

Rocky Scopelliti

Published in Australia by
Youthquake
E-mail: RockyS@Youthquake4.com
Website: www.youthquake4.com

First published in Australia 2020
© Rocky Scopelliti 2020

National Library of Australia Cataloguing-in-Publication entry

A catalogue record for this
book is available from the
NATIONAL
LIBRARY   National Library of Australia
OF AUSTRALIA

ISBN: 978-0-646-82154-2 (paperback)
ISBN: 978-0-646-98783-5 (ebook)

Printed by Ingram Spark

AUTHOR'S NOTE

It's often said that hindsight is 2020. So, as the third decade of the new millennium kicks off, I'll be putting that wisdom to the test. With the benefit of 20/20 hindsight (in 2020), and with ground-breaking new research, Australia 2030, which has been augmented with commentary from leading thought leaders, investigates Australian professionals' attitudes towards the coming decade. These attitudes were formed before and during a time when the nation's and the world's worst crisis since the last world war is unfolding.

The question of how to increase our capacity to adapt to a world of accelerated change has been thrust upon all of us by COVID-19 and it will be the test of that wisdom that will define the society we will become in 2030. The central questions we are all searching for answers to are what to believe – who, what, how and where the bloody hell are we? which road do we take? and what is on the other side anyway?

DEDICATION

To my beautiful children.

Believe in miracles because there's one in the form of a pure idea in each of your minds – just waiting to be acted on. Your abundant curiosity of the world and endless imagination will lead you to the treasure.

CONTENTS

Introduction

"When it comes to our ecological, social, cultural and economic future, misplaced optimism is as dangerous as blind faith. What is needed is the courage to face the way things are, and the wisdom and imagination – informed by the best available evidence – to work out how to make things better[1]"

~ Dr Hugh Mackay, social researcher

A word cloud is an informative image that thematically analyses and communicates much in a single glance. This word cloud reflects what you'll experience in reading this book.

Source: Australia 2030 research Rocky Scopelliti

As Australia entered the last decade, Hugh Mackay, arguably Australia's most respected and perceptive social researcher, and for me personally, the inspiration for my life's thought leadership research, took a long hard look at our society in the 21st century. Mackay published his findings in 2007 in a book titled *Advance Australia - Where?* He proposed that "while we enjoy unprecedented levels of prosperity and the promise of more to come, we are still battling an epidemic of anxiety and depression, taking on record levels of debt, and yearning for a deeper sense of meaning in our lives". At that time, Mackay challenged us to ask ourselves some hard questions:

- Can we improve our political system?
- Are we serious about global warming and renewable energy?
- Aren't we over the monarchy yet?
- Do we really believe in public education?
- Is poverty a fact of life, or a problem to be solved?
- Are we missing the golden opportunity by underfunding the arts?

As we now enter the next decade, these questions have become even more important to revisit and reflect upon. In fact, we should ask ourselves — did we really answer any of them?

From an economic perspective, Australia finished this past decade with an unprecedented economic cycle characterised by:

- Record low interest rates
- Record levels of debt
- Low inflation
- Low unemployment
- Booming exports
- Huge government spending.

What made that cycle unprecedented, is the fact that all these conditions have occurred at the same time. Real household income has been stagnant since the Global Financial Crisis (GFC), despite the fact that our national output has been rising. Combating that with record low interest rates was just not working and the rates runway does not have much left. We have never seen or experienced this cluster of those conditions before. And so we enter this new decade hoping it will be profoundly different to that of the decade just past.

A slowing economy means we tighten our belts. This nervousness manifests itself into a tightening of disposable income by our concerns for job security, the housing market, and stagnant wage growth just to name a few. Australian professionals today feel worse off than they did in the past decade. Progressive interest rate cuts have reinforced the perception of uncertainty not just to Australian professionals, but to the broader community. At we began 2020, these conditions were threatening the economy with recession over the coming decade while our dependency remains on the housing market.

Sadly, the impact of the drought and bushfires that closed the past decade and opened the new for Australians, has left us with many unanswered questions. The unprecedented impact of those bushfires on us all directly or indirectly has left us wondering about our future. It's left us knowing we have decisions to make and raised our concerns about the choices available to us.

Then came COVID-19 – The month we will never forget.

So, if the past decade didn't create enough uncertainty, the start of this new decade surely has. The emerging impact of the COVID-19 pandemic has sent fear around the world and is predicted to have a much worse impact than the GFC on global markets and the Australian economy. Unprecedented quantitative easing measures have been announced by the Reserve Bank of Australia (RBA) and Federal Reserve Banks around the world to battle the economic impact of the pandemic.

What began as a health crisis rapidly became an economic, social, cultural and scientific crisis also. Seemingly entrenched modes of orchestrating government were thrown away by COVID-19. For example, the decision to scrap the Council of Australian Governments (COAG) that was established in 1992, and simply keep the National Cabinet going was made in weeks and with remarkably little political bickering. To stave off a recession, on 12 March 2020, the Australian government announced an initial $17.6 billion economic stimulus package[2]. Then in unprecedented emergency actions, the RBA Governor Philip Lowe announced the cut in interest rates to a further 0.25 per cent in an out of cycle meeting to curb the economic impact from the COVID-19 pandemic. The RBA announced that it will inject $90 billion of ultra-cheap loans into commercial banks to on-lend so that banks can support their customers and risks of international credit markets tightening. On 30 March 2020, the government announced the $130 billion Job Keeper Payment to help keep Australians in jobs as we deal with the significant economic impact from COVID-19. This brings the government's total support for the economy to $320 billion across the forward estimates, representing 16.4 per cent of Australia's annual GDP3. Quite simply the need to quickly make policy decisions and get announcements out to bolster confidence overrode the usual layers of checking and rechecking. For example, the bill for the government's economic rescue package was subsequently, and dramatically, cut by $60 billion due to a miscalculation by Treasury[4].

National emergencies were announced worldwide to 'get ahead of the curve' as the world held its breath for a vaccine and treatments. Measures included banning travel, closing borders, banning the gathering of large crowds, social distancing, social isolation, closing schools, evacuation of staff in buildings, suburban lockdowns and quarantining (See Exhibit 1.1).

Exhibit: 1.1 Major COVID-19 developments between 12–21 March 2020

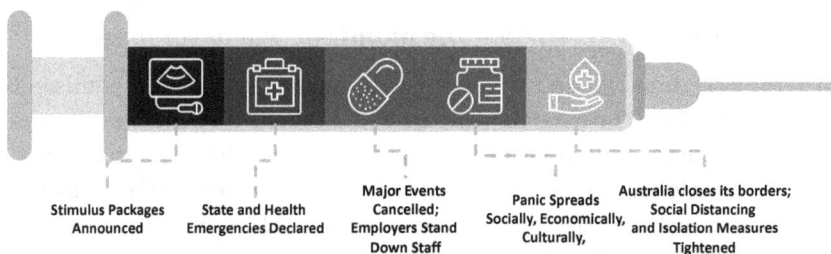

| Stimulus Packages Announced | State and Health Emergencies Declared | Major Events Cancelled; Employers Stand Down Staff | Panic Spreads Socially, Economically, Culturally, | Australia closes its borders; Social Distancing and Isolation Measures Tightened |

Ministers and prime ministers succumbed to the virus highlighting that no one was immune to this highly contagious disease. Talks of nationalising organisations that fell into financial difficulty surfaced as Virgin Australia went into administration and many others faced the same risk. Amid the COVID-19 pandemic, many people transitioned to working—and socialising—from home. For example, Zoom Communications, a video conferencing company, reported daily meeting participation rises from 10 million in December 2019, to 300 million in April 2020. That growth saw its market capitalisation skyrocket by 129 per cent over that three-month period to US$48.8 billion. To put that in perspective, Zoom's market capitalisation became more than that of the seven biggest airlines in the world whose combined market capitalisation collapsed to US$46.2 billion over that same corresponding period[5].

Just as the catastrophic bush fires over summer 2019/20 were extinguished with welcome rainfall, another catastrophic event of much greater global impact hit our shores—COVID-19. Its impact has struck fear and uncertainty around the world impacting people's lives and work, global economies, major industries, healthcare systems and much more. This infectious disease impacted all facets of our society with schools closing, friends and family stuck in overseas countries, the closing of borders and organisations pushing workforces into their homes to work remotely. Shops closing, mass employment layoffs, police patrolling the streets enforcing social distancing, beaches closed, sports and entertainment banned, supermarkets raided for essential goods – within two weeks, our whole way of living, working and playing stopped.

One massive societal amygdala hijack

The fear this event has created on a global scale has not been seen before. Our vulnerability to disease has manifested itself perhaps in what some might describe as undesirable and unusual social behaviours. These included the rush on buying toilet paper despite the fact the vast majority of it is produced locally and the emergence of racial tensions targeting those nations where the virus became prevalent early. We seem to have experienced one massive societal amygdala hijack with people seemingly reverting to primal survival instincts, pausing their common sense, values and the spirit of helping one another. The speed, scale and impact of this virus has left us knowing that not all decisions and choices people make follow an orderly, predicable pathway.

COVID-19 and our responses to it, highlighted a range of societal complications and issues associated with the significant demographic changes occurring in Australia. Australia is at a demographic tipping point.

At one end, the proportion of elderly are increasing. Baby Boomers (broadly 54–72 years) that are now turning 65 years of age and retiring, are predicted to cost the Australian Government $36 billion per year by 2028. That's more than Medicare today. By 2057, 22 per cent of the Australian population are predicted to not be contributing to the economy, from 15 per cent today[6]. It is the elderly who are the highest risk of severe effects from COVID-19, and hence in greatest need of social distancing – but they are also already amongst our most socially isolated people.

At the other end we see the rise of Millennials (broadly 19-39 years). As consumers, employees, investors and policy makers, Millennials will permeate all facets of our society. However, many Millennials face the prospect of long-term unemployment resulting from potentially the first recession they have ever experienced. What impact will (potentially) years of COVID-19 related uncertainty has on their worldview?

How can Australia increase its capacity to adapt to a world of accelerated change? What road will we choose?

This question has been the subject of my research for more than 16 years now. So, what is the answer? Well unlike a closed puzzle where we know what answer will look like before we begin, or an open puzzle where we need to figure out the answer as we go, this question behaves more like a mystery that keeps us on the edge of our seats and induces our curiosity every step of the way, and just when we think we know the answer, another mystery emerges. The unprecedented events of the bushfires over the summer and the twists and turns of COVID-19: the measures we've taken to control it, to respond to its impacts and how our reactions will unfold at an individual, demographic and societal level over years to come are a fine example of the mystery. For the world and for Australia, the setting for this mystery is the 4th Industrial Revolution that is impacting every nation, industry, economy, organization, society and, individual. It is a mystery that will dictate our social, cultural, economic and technological future.

Like the plot of any good mystery, our journey forward will undoubtedly feature heroes and villains, points of tension and struggles sometimes verging of hopelessness. It will feature critical crossroads where the wrong path is chosen points. It may even feature moments of farce. But ultimately there is also much to look forward to in the resolution – improved health, education, lifestyles, jobs new businesses and industries and the potential to help the emerging generations who are our future to be better humans than we.

But that optimism, is causing fear about change. For example, the words 'robotics' or 'artificial intelligence' translate into the message of 'job loss' on the streets. Technology such as mobile apps that monitor our contact exposure to COVID-19, translate into intrusion of privacy and present cyber risk. This mystery will be unlike any other because of the significant advancements of the technology revolution, and the willingness of our society to tackle past unsolved problems and unexplored horizons.

Reflection

So, the mystery we must also address about the coming decade is:

Without the perceived safety net of a predictable, linear future, which road will we choose? And where do we think we will arrive at?

For Australia, the past decade saw transformational change that has altered many aspects of our society. The speed, scale and impact of it was unlike any other decade in our history. Therefore, it's critical for us to pause, reflect and ask the question:

Where the bloody hell are we?

- How did we get here?
- Is this where we expected or wanted it to be?
- How do we feel about where we are?
- Where to from here?
- What road should we take for the coming decade?
- Who's coming?

What's on the other side?

Exploring those questions and many more, are what you're about to experience in this book. That experience will centre on the collision of eight megatrends that collectively will define Australia politically, economically, environmentally, regionally, socially, trustworthily, knowledgably, scientifically, and technologically over the coming decade.

This book is based on an investigation into the attitudes of Australian professionals and our leaders to the decade ahead. By invitation, 673 Australian professionals participated in the quantitative and qualitative study conducted during January–April 2020 that will be from here referred to as the Australia 2030 research. It considered our attitudes towards the decade ahead, including the following questions:

- How confident are we in the government's plans for our future?
- What issues do we predict will affect Australia and the world?
- What qualities do we expect from our leaders?
- What contribution do we want organisations to make in our society?
- What is our place in the world and region?
- What are our concerns about our jobs and the workplace?
- Who do we trust to control our best interests?

- How do we feel about the impact of technological & scientific developments on our personal, professional and family lives?
- Are we optimistic or pessimistic about technological & scientific developments?

While our attitudes are not predictors of our future behaviour, they do none-the-less reflect how we have reacted to political, economic, technological, demographic, health and social developments and issues that manifested themselves in the past decade and beyond. Choices and pathways that seemed difficult for us in the past decade may, for better or worse, seem clear. For others the reverse may be true. While is easy to know the right thing to do after the fact, it's much harder to predict and prepare for the future. This is of course the benefit of 20/20 hindsight – and in this case 2020 hindsight! For 20/20 hindsight is not really intelligence. The wisdom to explore, prepare for and adapt to a world of accelerated change based on a considered understanding of the society we aspire to create most definitely is intelligence. Importantly however, today's attitudes provide a platform to consider what roads we may choose, or not. They do play a crucial part in this decision making through a complex tapestry of global influence and generational change.

New leadership thinking

According to the World Economic Forum, we have much work to do. Work that must be done collaboratively and globally if it is to benefit everyone with collective leadership of both public and private sectors. It terms this leadership as 'systems leadership', described as cultivating a shared vision for change, collaborating with all stakeholders of a global society, and executing to garner system benefits[7]. Large-scale initiatives are often driven and supported by people who fit a certain profile – those who are able to catalyse and empower collective action among others, rather than controlling or directing the action themselves. These people are increasingly described as systems leaders. The systems leadership (technological, governance, values etc) it describes is not targeted to government or business leaders, but rather a paradigm that empowers all citizens and organisations to invest, innovate and deliver value.

"We must develop a comprehensive and globally shared view of how technology is affecting our lives and reshaping our economic, social, cultural and human environments. There has never been a time of greater promise, or greater peril."

~ Klaus Schwab, founder and executive chairman, World Economic Forum

Where the bloody hell are we? Do we take 'the road less travelled', or 'the road not taken'? and 'what's on the other side anyway'?

This is the crossroads we now find ourselves in. Never before, have we asked those questions during a time when that all aspects of our society have become impacted. What is our place in the world? The greatest opportunity and responsibility we all now have as we commence this new decade, is to answer these questions. For it is those answers that will ultimately define our future this decade, our place in world and the indeed, the characteristics of the world we bequeath to our children. These are not decisions resting in the hands of few but must rather include every citizen. Answering those questions, will also require us to reflect critically on the decade just past, and decide which aspects of it we should continue building on, and which we should ensure are not repeated.

The Road Not Taken is an ambiguous poem by Robert Frost that invites its readers to think about choices in life – do we go with the flow or do we make our own way? If life is a journey, this poem highlights those times in life when decisions need to be made. Which way will you go? Which road will you take?

The ambiguity springs from the question of free will versus determinism. Does the speaker in the poem consciously decides to take the road that is off the beaten track or only does so because the road with the bend in it isn't appealing? External factors therefore make up his mind for him.

It's said that Robert Frost wrote this poem to feature a trait of his friend Edward Thomas, an English-Welsh poet, who, when out walking with Frost in England would often regret not having taken a different path. Thomas would often regret not taking another path that might potentially have offered better opportunities, despite the outcome being unknown.

Frost told Thomas: "No matter which road you take, you'll always wish you'd taken another." Ironically, Frost meant the poem to be a somewhat light-hearted critique, but it has turned out to be anything but. People take the philosophical conundrums it highlights very seriously.

It is the hallmark of the true poet to take such everyday realities, in this case, the sighs of a friend on a country walk, and transform them into something so profound. *The Road Not Taken* is really about what did not happen: This person, faced with an important conscious decision, chose the least popular path, the path of most resistance. He was destined to go down one, regretted not being able to take both, so he sacrificed one for the other.

In the end, readers are left to make up their own mind about this mystery. Was the choice of the road less travelled a positive one? What would have happened had he chosen the *'road not taken'*? In 2030, how will we reflect on the decade just passed? How will we judge ourselves on the choices we made? How will you judge yourself on the choices you made? Did you take the right road? What would have happened had you chosen a different road?

The main theme of this book as with the poem "The Road Not Taken", is that it is often impossible to see where a life-altering decision will lead. For example, in 2010, we simply could not have known the cultural social, economic and political impact of the choices we made as a nation or individually. The future after all, is not there to be predicted. It does not cooperate with our prognostications. However, our natural curiosity will leave us wondering what the outcome would have been if the other road, *the road not taken*, was the road chosen. This can only ever be a 'hypothetical' as it is impossible to say whether taking the other road would have left us better or worse: all we can reliably assume is that it would have been different.

This poem *'The Road Not Taken,'* like a great mystery, presents us with a dilemma. Which road do we choose? Like the poem, the central message of this book is that, in 2020 as is with life, we are presented with choices – we are required to make a decision. Viewing a choice as a crossroad, it becomes clear that we must choose one direction or another, but we can't choose both.

In this book, our retrospective look at the past decade is not to determine whether the road we chose was the right road or the wrong road, whether it was *the road not taken*, or *the road less travelled*, but rather, an account of the events and developments that has led us to where we are today in 2020, and the attitude's and beliefs we have formed. It is those which will inform our decisions on the journey forward.

This book is not about encouraging us to take one road over another, about our individuality or our uniqueness as a society. For it is our curiosity to contemplate the "What if..." scenarios about the choice we did not make and hypothesise about the outcomes that might have achieved. After all, that is the beauty of 20/20 hindsight. This pondering about the different life we may have lived had we done something differently is central to *The Road Not Taken*. But what makes this coming decade unique, is that for so many of the issues before us such as politics, the environment, the economy, changing global power, trust, our role in Asia, technological and scientific developments such as gene editing, artificial intelligence and many more, we will not be able to return and try the 'original' road again. This decade will present us with many choices – perhaps some more profound than many decades before. But for many of those, going backwards will no longer an option.

Life is anything but usual and uncertainty has become the new normal as we transition into the 4th Industrial Revolution.

We are living in the age of accelerated transformation. While transformation itself is not necessarily new to us, the frequency and pace of change is higher, and time-to-impact of our decisions is shorter than it has ever been. While our traditional linear view of the future, the models and methodologies we use to forecast, create policy upon, operate our businesses and the associated legacy technologies may have served us well in the first, second and part of the 3rd Industrial Revolutions, they are inadequate and unreliable predictors of the future needs of societies, value creation, behaviour of markets, and economic performance or survival in the 4th Industrial Revolution.

This revolution is characterised by emerging technology breakthroughs with potentially highly disruptive effects. Technologies such as artificial intelligence, robotics, the Internet of Things (IoT), autonomous vehicles, virtual and mixed reality, 3D printing, biotechnology, nanotechnology, materials science, energy storage, blockchain, 5th generation mobile networks (5G), quantum computing and many more.

This traditional 'road taken' has been a linear view of the world as we have sought to control, even predict the future. However, we are now living in a world characterised by exponential change where the end of the road is no longer a distant horizon with a well-defined road to it, but rather, the corner block 100 metres away, where we don't know what is around the corner, let alone whether we need the pavement, roadway, motorway, or airspace that we will need to travel on to get there.

COVID-19 introduced the concept of 'exponential' systems to the mass market. Getting ahead of 'the curve' and 'flattening the curve' became daily targets to control this highly contagious virus. Nations were now being compared to one another on the exponential curves and nations such as Australia, New Zealand and Singapore were being hailed for the initiatives put in place to flatten the exponential curve. The curve has now become the symbol of the speed, scale and impact of change that is our world today. Importantly, it is the symbol of how we need to increase our capacity to adapt to that world.

As Klaus Schwab, founder and executive chairman of the World Economic Forum, describes "we are at the beginning of a 4th Industrial Revolution that is fundamentally changing the way we live, work and relate to one another". He also proposes that "businesses, industries and corporations will face continuous Darwinian pressure and as such, the philosophy of "always in beta" always evolving will become more prevalent"[8].

In his 30 years of research, Ray Kurzweil[9], a world-leading author, computer scientist, inventor, futurist and co-founder of the Singularity University, has made key observations that explain this exponential environment. While we cannot predict the future, Moore's law and Kurzweil's law of accelerated returns allow us to predict the pace of change. Gordon Moore observed that the number of transistors on a silicon die (which is one measure of computing power) doubles every 18 months – a classic case of exponential growth. This observation is known as "Moore's law" Kurzweil proposed the law of accelerating returns, which states that the rate of change in a wide variety of evolutionary systems (including but not limited to the growth of technologies) tends to increase exponentially10. Kurzweil argued for extending Moore's law to describe exponential growth of diverse forms of technological progress. Whenever a technology approaches some kind of a barrier, according to Kurzweil, a new technology will be invented to allow us to cross that barrier. He predicts that such paradigm shifts have and will continue to become increasingly common, leading to "technological change so rapid and profound it represents a rupture in the fabric of human history."

This philosophy directly informs the question at hand – if uncertainty is the new normal for the coming decade, which road will we choose?

To address that question, we must understand the era into which this next decade will unfold: the imagination age. The information age, where analysis and thinking were the main activities driving economic growth, is widely held to have begun in the 3rd Industrial Revolution. The imagination age, where creativity and imagination will become the primary creators of economic value, is posited to heralded by the 4th Industrial Revolution – a revolution which is already underway. The concept holds that technologies and scientific developments of the 4th Industrial Revolution, such as artificial intelligence, nanotechnology, biotechnology, material science, space, robotics, augmented, virtual reality and many others, will change the way humans live, work and relate to one another, as well as our economic and social structures. This will raise the value of imagination work of creativity, designers, architects, artists, producers over rational thinking as a foundation of culture and economics.

The term imagination age was first introduced in an essay by designer and writer Charlie Magee in 199311. Magee proposes that the most successful groups throughout human history have had one thing in common: when compared to their competition they had the best systems of communication. The fittest communicators—whether tribe, city state, kingdom, corporation, or nation—had (1) a larger percentage of people with (2) access to (3) higher quality information, (4) a greater ability to transform that information into knowledge and action, (5) and more freedom to communicate that new knowledge to the other members of their group. This idea may well explain why some countries got 'ahead of the COVID-19 curve' and others didn't – same tense.

So, can Australia's capacity to imagine as defined by Magee explain why with systems leadership, Australia used that imagination to adapt socially, economically, technologically, and culturally to get ahead of the COVID-19 curve? If so, then we should welcome with optimism the decade ahead that will see Australia play a significant role in the augmentation of our digital, physical, biological and environmental worlds that is the 4th Industrial Revolution.

The insights in this book are for individuals, leaders and policy makers seeking to shape Australia's future, your future, and that of our loved ones. Its purpose is not to provide you with the answers. After all that would spoil the mystery of the decade, ahead right. Rather, it is intended to provide you with clues and thinking to help understand the signposts along the way.

As Dr Hugh Mackay so wisely said *'What is needed is the courage to face the way things are, and the wisdom and imagination – informed by the best available evidence – to work out how to make things better.'*

Importantly, I hope these insights inspire you to consider the role you can or want to play in making the choices of which road we will take. For individuals, choosing the road less travelled might just be one of your most rewarding life experiences.

Let's put our imagination to good use.

"Imagination is everything. It is the preview of life's coming attractions.
Only those who believe that anything is possible,
can achieve the things, most would consider impossible"

~ Albert Einstein

CHAPTER 1

Politics

"That's not a knife...This is a knife."
(Crocodile Dundee)

'When politicians offer you something for nothing, or something that sounds
too good to be true, it's always worth taking a careful second look.'

~ Malcolm Turnbull, Australian Prime Minister, 2015–2018

This word cloud thematically analyses and presents a creative reflection of
how we felt about Australian politics over the past decade.

Morrison report Senate Court
decade major Australia budget program
bloody government Confidence Abbott Trust
Tax crisis politics period leaders Fraser
Commission boats year Scandals
resigned Minister Labor ministerial
parliament shirtfront While controversy Gough
campaign backflip Robodebt passed zip time Speaker Federal
crisis former election Party Eligibility Community
faceless Gillard policies Tony bills Grangegate Pink Batts
funding funds leader Rudd Choppergate Canberra emerged
Liberal Credlin Malcolm April Sorry Scott Whitlam
voters Treasurer Rorts trust Slipper Turnbull
Peter Sports Australian
leadership Bob displaced political
electorate miracles parties hell
Kevin legislation men Offshore detention
policy displaced HSU allegations chaos Gough
power carbon

Source: Australia 2030 research Rocky Scopelliti

This chapter will reflect on the major political events and key policies throughout the period between 2010–2020 to consider how they may have shaped and influenced our attitudes. It may provide further context as to why we may hold the views that we do today and about trust and confidence with government over the coming decade.

Fourteen years ago, an Australian 18-year-old teenage model from Cronulla in New South Wales was catapulted into the limelight when she broke out with a smile, on a beautiful white sand beach, and uttered six words that went on to create a national, and indeed international uproar. That person was Lara Bingle (now Worthington) and those words were "Where the bloody hell are you?". This was the tagline to invite visitors to our shores in a reported $180 million advertising campaign by Tourism Australia. The campaign's objective was to attract visitors from Japan, Germany and the United Kingdom to Australia. Although the campaign reportedly fell well short of those objectives, it became one of our most memorable tourism taglines since "I'll slip an extra shrimp on the barbie for ya" by Paul Hogan. Banned in the UK amid fierce debate, the 'Where the bloody hell are you?' commercial was slammed for its perceived vulgarity. However, while delivering a keynote speech during a visit to Canberra, Tony Blair (former Prime Minister of the United Kingdom) said his first thought on arrival in Australia was "Where the bloody hell am I?"

In December 2019, that tagline resurfaced through a social media frenzy but for a different type of invitation. This time it targeted the Australian Prime Minister Scott Morrison in a call to return to Australia from his holidays in Hawaii and lead the country through a bushfire catastrophe of unprecedented global scale and impact. Lara Worthington in a tweet, asked the Prime Minister 'WHERE THE BLOODY HELL ARE YOU???' (see Exhibit 1.1)

Exhibit 1.1: Tweet from Lara Worthington

Lara Worthington ✅
@MsLWorthington

Scott Morrison : WHERE THE BLOODY HELL ARE YOU???
#AustraliaBurns #AustraliaFires

♡ 4,942 3:32 PM - Dec 18, 2019

💬 1,467 people are talking about this

Source: Twitter

Ironically, Scott Morrison was the managing director of Tourism Australia in 2006 when the "where the bloody hell are you?" advertising campaign was launched. But getting called out by the woman he made famous, who was originally from his own electorate, the seat of Cook, may well be the 'karma' he will never forget.

Over the ensuing months, the tag line remained in the media locally and internationally and was used to call out other ministers, who were absent, hypercritical, evasive or just plain ignorant to the scientific facts on climate change while this catastrophe unfolded around them. The tagline quintessentially captured the sentiment of disappointment in the lack of political leadership among Australians. Climate change is one of the topics where lack of alignment between policy, the espoused position of political leaders and the position of the general populate is most apparent. However, this anger did not surface simply as a response to that single tragic event, but rather reflects the culmination of growing public frustration with political misalignment, inaction, intransigence and shenanigans over the last decade.

Political leadership 2010–2020

'In politics, if you want loyalty, get a dog.

~ *Peter Costello, federal government treasurer 1996–2007*

In 2020, Australia closed one of the most politically unstable periods in its history. Our nation was divided, disillusioned and disappointed by the political 'self-indulgence' demonstrated by politicians from all political parties and independents included. As unexpectedly spectacular as the end of the decade was politically for Prime Minister Scott Morrison's leadership ("I've always believed in miracles"), so too it was for Kevin Rudd who commenced the decade as Australian Prime Minister, and in his farewell speech after being displaced by Julie Gillard departed with the phrase "we've [Therese and Kevin] got to zip"[12] .

Between 'zipping' and 'miracles', this past decade saw the prime ministerial leadership change five times within an eight-year period. The electorate became spectators to a democratic system where their best interests took a back seat to factions and political self-interest. The resulting political policy backflips throughout that period were no longer became fodder for opposition parties to challenge, as all parties feasted on them abundantly. It was as if all sides of politics believed they possessed infinite social capital and an endless supply of goodwill. For the electorate this meant that voting for a party or leader espousing particular policy came with less and less assurance that policy would be progressed.

The Australian National University's (ANU) Australian Election Study[13] shows a majority of voters thought the removal of the leader was unjustified. We became sensitised to these leadership spill announcements as both parties delivered more policy uncertainty. That uncertainty has resulted in Australians no longer understanding what each party truly stands for or is committed to. Policies became seasonal features of the electoral cycle, but much less appetising than a good winter navel orange.

More than any other time in our history, this past decade has led to greater disengagement, lower tolerance, less trust and, perhaps, lower expectations of politicians themselves, their policies and the major parties they represent.

'Faceless Men'

"Valar Dohaeris. All men must serve. Faceless Men most of all."

~ Jaqen H'ghar, Game of Thrones

In the film series Game of Thrones, the Faceless Men are a group of assassins based in the city of Braavos. They consider themselves to be servants of the 'many-faced god' and possess the ability to physically change their faces, shapeshifting such that they appear as an entirely new person. In the series, the Faceless Men only assassinate targets they've been contracted to kill.

The term 'faceless men' surfaced again in the past decade of the Australian political vernacular to describe those who conspired to bring down the standing prime minister at the beginning of 2010 – Kevin Rudd. Senators David Feeney, Mark Arbib and Don Farrell along with then Victorian MP Bill Shorten, were reported to have initiated the moves that resulted in the replacement of Kevin Rudd by Julia Gillard as Australia's Prime Minister and leader of the Australian Labor Party. Then Rudd was returned displacing Gillard, but this time Rudd tore apart more than 100 years of Labor Party tradition to remove the power of the faceless men who he felt had knifed him in 2010. Instead, the power to remove and install a new leader would be in the hands of rank-and-file members of the party, making it nearly impossible for a Labor Party leader to be removed. Those changes would likely have stopped Rudd from removing Julia Gillard as leader two weeks prior to these changes and stopped her from ousting him in 2010. Rudd explained "You want to be able to say to the Australian people, you vote for this guy, you vote for this woman, they end up staying on for the duration of the term"[14].

But the faceless men were not just a cohort unique to the Labor Party. They reappeared again in 2015 when Tony Abbott, the standing Liberal National Party leader and Prime Minister was brought down by Malcolm Turnbull. Scott Ryan, Wyatt Roy, Mitch Fifield, Mal Brough and Peter Hendy, were reportedly, the faceless men who conspired against Abbott[15].

Then came Turnbull's turn. With members of the Liberal National Party believing Turnbull could not win the next election, the faceless men and women made their move. Peter Dutton challenged Turnbull with support from Mathias Cormann, Mitch Fifield and Michaelia Cash, who delivered the coup de grace by confirming publicly they had abandoned the prime minister. Interestingly, James McGrath, the Queensland senator, who masterminded the campaign to displace Abbott in 2015, emerged from the shadows as the 'faceless man' that provided the numbers to displace Turnbull. While Dutton's challenge for the Liberal party leadership was unsuccessful (40-45 votes), Scott Morrison emerged as the new prime minister. A tweet from former prime minister Julia Gillard, who began the decade of political leadership instability, and subsequently became a victim herself, welcomed the new prime minster Scott Morrison and shared advice for the displaced Malcolm Turnbull on life after politics. Perhaps also, some early advice to Scott Morrison in the event that he too is displaced by faceless men looking to install a new 31st prime minister of Australia (see Exhibit 1.2).

Exhibit 1.2: Tweet from former Prime Minister Julia Gillard

Julia Gillard ✓
@JuliaGillard

Congratulations to @ScottMorrisonMP - always an honour to serve. To the 29th PM, from the 27th PM, @TurnbullMalcolm there is a life after. Best wishes for the days to come.   Exhibit 1.1

4:29 pm · 24 Aug 2018 · Twitter for iPhone

1.7K Retweets   **15.2K** Likes

Source: Twitter

The 'faceless men and women' set the scene for an unprecedented decade in our political history where so many prime ministers were removed from power by their own parties. No other country changed leaders more often than Australia did in the eight-year decade— Kevin Rudd (2010), Julia Gillard (2013), Kevin Rudd (2013), Tony Abbott (2015) and Malcolm Turnbull (2018) and Scott Morrison (2018 -) (see Exhibit 1.3).

Exhibit 1.3: Australian prime ministers (2010–2020)

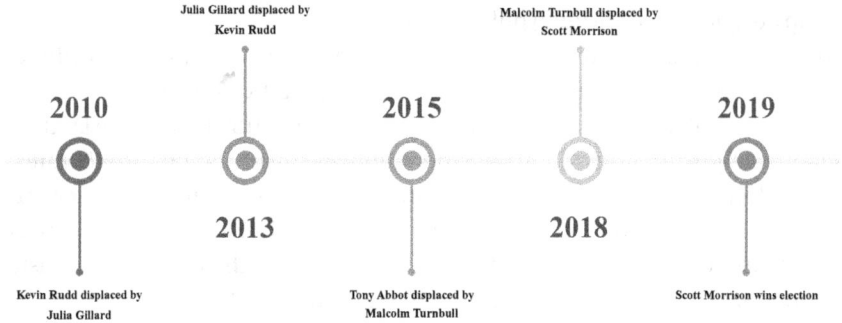

Julia Gillard displaced by
Kevin Rudd

Malcolm Turnbull displaced by
Scott Morrison

2010   2015   2019

2013   2018

Kevin Rudd displaced by
Julia Gillard

Tony Abbot displaced by
Malcolm Turnbull

Scott Morrison wins election

The prime ministerial office became a revolving door of leaders chosen by their many-headed gods and executed obediently by their faceless men. The subsequent animosity between the anointed and displaced leaders, such as Rudd & Gillard and between Abbott & Turnbull became a public spectre destabilising major Australian political parties, and their policies.

Yet as we commence the new decade, there are no signs that the revolving door of party leadership days is isolated to our past decade. The first sitting day in 2020 for federal parliament, which was to be dedicated to offering condolences to bushfire victims, was instead overshadowed by another Liberal National Party leadership spill. This time in the National Party between Michael McCormack, the standing leader, and Barnaby Joyce, the former leader. McCormack won the spill and consistent with other displaced leaders, Joyce remains on the sidelines throwing stones at his own party.

The past decade showed us just how big the gaps are between factions of major political parties. The manner with which these differences escalated and in the way they were reconciled fell well short of community expectations from its major political parties.

New populist parties emerged to attempt to capture the growing pool of disillusioned voters. Most notably Pauline Hanson's One Nation, Clive Palmer's United Australia Party and Bob Katter's Australian Party. The Greens led by Richard Di Natale at that time had become the most successful minor party receiving 1.4 million lower house votes at the May 2019 general election, which represents a 10.4 per cent share.

Political scandals and controversies

Not only did this past decade see a world record of prime ministerial leadership changes, but more political scandals and controversies from both state and federal governments than any other decade in our political history (see Exhibit 1.4).

Exhibit 1.4: Major Australian federal and state government scandals and controversies (2010 – 2020)

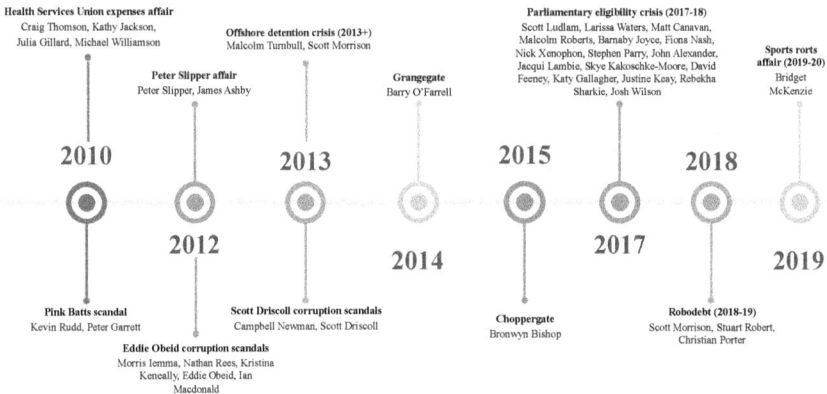

Health Services Union expenses affair
Craig Thomson, Kathy Jackson, Julia Gillard, Michael Williamson

Peter Slipper affair
Peter Slipper, James Ashby

Offshore detention crisis (2013+)
Malcolm Turnbull, Scott Morrison

Grangegate
Barry O'Farrell

Parliamentary eligibility crisis (2017-18)
Scott Ludlam, Larissa Waters, Matt Canavan, Malcolm Roberts, Barnaby Joyce, Fiona Nash, Nick Xenophon, Stephen Parry, John Alexander, Jacqui Lambie, Skye Kakoschke-Moore, David Feeney, Katy Gallagher, Justine Keay, Rebekha Sharkie, Josh Wilson

Sports rorts affair (2019-20)
Bridget McKenzie

2010   2013   2015   2018

2012   2014   2017   2019

Pink Batts scandal
Kevin Rudd, Peter Garrett

Eddie Obeid corruption scandals
Morris Iemma, Nathan Rees, Kristina Keneally, Eddie Obeid, Ian Macdonald

Scott Driscoll corruption scandals
Campbell Newman, Scott Driscoll

Choppergate
Bronwyn Bishop

Robodebt (2018-19)
Scott Morrison, Stuart Robert, Christian Porter

Some of the major ones in chronological order were:

2010

Pink Batts scandal. Implemented by Minister Peter Garrett in the Rudd government as part of its Energy Efficient Homes Package, the Home Insulation Program became the subject of controversy by the government's under-estimation of the risks involved resulting in the deaths of four workers from independent incidents. The program attracted a raft of inexperienced, poorly trained and poorly overseen installers. Queensland coroner Michael Barnes found the risk of electrocution "was not appreciated" by the government authorities from the outset. Allegations of fraud surfaced from businesses and individuals in New South Wales, Queensland and Victoria. The Home

Insulation Program was estimated to have cost taxpayers $1 billion with an additional $1-$1.5 billion required to remediate installation problems[16].

A Royal Commission was established in 2013 overseen by a sole Royal Commissioner, Ian Hanger AM QC, to investigate the deaths, workplace occupational health and safety and other risks associated with the program. Kevin Rudd appeared before the commission becoming the first prime minister to have been called to appear before a Royal Commission. Peter Garrett also appeared before the Royal Commission and claimed ultimate responsibility for the implementation of the program.

Health Services Union (HSU) expenses affair. This scandal concerned criminal activities by Craig Thomson, a former national secretary of the HSU. At the time of the allegations, he was serving as a minister in the Gillard government. In 2014 Thomson was found guilty of 13 charges of theft and convicted for using HSU funds for personal use and fined $25,000. He was sentenced to 12 months' imprisonment, with the conviction and sentence later overturned on appeal.

In 2013, Michael Williamson, a former HSU national president, pleaded guilty to two charges of fraud involving $1 million, fabrication of invoices and hindering police investigations. Williamson was sentenced to seven and a half years imprisonment.

In 2015, Kathy Jackson, the national secretary of the HSU was found to have misappropriated HSU union funds by the Australian Federal Court in a civil case brought against her by HSU and ordered to repay $1.4 million in compensation. The court found that she had used HSU funds to buy personal items for herself, including through cash withdrawals, misusing three credit cards and funds from a payment made by a cancer hospital in Melbourne to settle a back-pay dispute[17].

2012

Peter Slipper affair. In 2011 Peter Slipper was nominated unopposed and installed as Speaker of the House of Representatives. His nomination by the Labor Party was considered controversial by the opposition leader Tony Abbott. Slipper resigned from the Liberal National Party and continued as Speaker as an independent representative. In April 2012, Slipper was accused of misusing Cabcharge vouchers which lead to an investigation by the Federal Police. Unrelated, Slipper was also accused of sexual harassment from a member of his staff, James Ashby. While these allegations were dismissed by the Federal Court that same year, Slipper stepped aside as Speaker.

In January 2013 the Federal Police alleged three offences relating to allegations of the misuse of taxpayer funded Cabcharge vouchers to pay for hire cars to visit a number of wineries. In 2014, Slipper was found guilty and sentenced to 300 hours community service and ordered to repay the $954 taxpayer funds. On appeal, Justice Burns, ruled the appeal be upheld and the conviction and sentence be set aside[18].

Eddie Obeid corruption scandals. In 2012 the Independent Commission Against Corruption began hearings relating to Obeid's interests in property and mining. The investigation considered the role of the Minister for Primary Industries and Minister for Mineral Resources Ian Macdonald in decisions made in 2008 to open a mining area for coal exploration. In July 2013 the Commission found that Obeid, Macdonald and others had engaged in corrupt activities in relation to the two mining sites. In May 2017, both were committed to stand trial. The crown alleges Obeid and Macdonald had a close working relationship and that Obeid stood to gain $60 million from the deal[19]. The case continues as I write.

2013

Offshore detention crisis. The opposition Liberal National Party led by Tony Abbott, campaigned heavily on asylum seekers with a policy slogan of 'stop the boats.' Following the election win by the Liberal National Party led by Tony Abbott, tougher polices were implemented to curb the stream of asylum seekers planning or heading towards Australia. Classifying the new policy 'Operation Sovereign Borders' under total secrecy allowed the government to avoid disclosure of the treatment of asylum seekers that were intercepted at sea.

In 2014, the Human Rights Commission published a damning report (The Forgotten Children)[20] of the treatment of children in immigration detention both onshore and on Christmas Island. While the Human Rights Commission was barred from visiting Manus Island and Nauru, the report was damning of the treatment of children in the custody of Australian Territories. In April 2016, the Supreme Court of Papua New Guinea ruled that detention of asylum seekers on Manus Island breached their constitutional right to freedom[21]. Despite that ruling, Immigration Minister Peter Dutton declared that the 850 people held in that facility, would not come to Australia[22]. Private military contractors were commissioned by the government to manage new centres at Lorengau.

After much controversy, in 2019, the amended legislation that would allow sick people to be treated in Australia, which had become known as "the Medevac Bill", passed in the House by 75 votes to 74 and passed in the Senate by 36 votes to 34. The *Home Affairs Legislation Amendment (Miscellaneous Measures) Act 2019*[23], now requires the approval of two doctors, but their approval may still be overridden by the Home Affairs Minister. Human rights advocates hailed the decision, with one calling it a "tipping point as a country", with the weight of public opinion believing that sick people need treatment.

2014

Grangegate. In April 2014, Barry O'Farrell, then the 43rd Premier of New South Wales, appeared as a witness during an investigation by the NSW Independent Commission Against Corruption into alleged actions by Australian Water Holdings (AWH). It was alleged that O'Farrell had received a $3,000 bottle of Grange Hermitage wine from an AWH executive, which he had failed to declare – allegations he denied. On 15 April, he was advised of a "thank you" note, to be presented to the ICAC, that he handwrote for AWH CEO Nick Di Girolamo. It read "Dear Nick & Jodie, We wanted to thank you for your kind note & the wonderful wine. 1959 was a very good year, even if it is getting even further away! Thanks for all your support. Kind regards."[24]

At a press conference on 16 April 2014, O'Farrell stated that he had had "a massive memory fail" and he still could not explain a gift that he had "no recollection of". He announced his intention to resign as the Premier of NSW that same day25. Although the Counsel assisting the Commission determined there was no suggestion O'Farrell engaged in corrupt conduct, ICAC's report subsequently cleared him finding that it was satisfied, "that there was no intention on O'Farrell's part to mislead".[26]

2015

Choppergate. In July 2015, Bronwyn Bishop, then Speaker of the House, became embroiled in the 'Choppergate' expenses scandal surrounding her use of parliamentary travel entitlements that ultimately led to her resignation as Speaker. She was found to have chartered helicopter flights from Melbourne to Geelong and back to attend a state Liberal Party fundraiser in November 2014. The cost of the flights was $5,227.27 for a journey that typically takes an hour each way by road. Bishop refused to resign over the expenses claim, describing it as an "error of judgement", while expressing disappointment that the controversy had become a distraction from the opposition and its policies[27] Later, she agreed to pay back the sum of the helicopter flight plus a penalty of $1,307.[28] Bishop resigned the Speakership in August 2015 and moved to the backbench.[29]

2017

Australian parliamentary eligibility crisis. In July 2017, a constitutional crisis emerged when 15 sitting Members of Parliament were ruled ineligible by the High Court of Australia or had resigned pre-emptively. Section 44(i) of the Australian Constitution, which prohibits parliamentarians from having allegiance to a foreign power, especially citizenship gave rise to the crisis particularly for those with dual citizenship.[30]

The resignation of John Alexander briefly cost the Liberal National Party Government its lower house majority, until Barnaby Joyce and Alexander regained their seats in by-elections. The crisis prompted calls for constitutional reform to prevent dual citizens from being disqualified, which would require a referendum.[31]

2018

Robodebt. In 2016, Centrelink began an automated data-matching process of welfare recipients' records against data from the Australian Taxation Office. The conversion from a human oversight-based process to an automated process increased the debt recovery letters issued per year from approximately 20,000 to 169,000.[32] Responsibility shifted from Centrelink needing to verify the information, to the individual to prove they did not owe funds. The new process was reported to save the government $300 million and if extended to the age pension and disability support pension, could yield as much as $1 billion.[33] In September 2019 Gordon Legal announced its intention for a class action suit challenging the legal foundations of the 'robo-debt' system.[34] The controversial system was ruled unlawful in 2019 with the Federal Court saying Centrelink could not have been satisfied the debt was correct. In a stunning reversal Government Services Minister Stuart Robert announced the government plan to refund more than $720 million to an estimated 373,000 people that were impacted by the system[35].

2020

Sports rorts affair. The title 'sports rorts' first emerged during 1993–1994 when the then Sports Minister Ros Kelly under the Keating Labor government resigned under controversy surrounding the awarding of sports funding.

In January 2020 the Australian National Audit Office published a report into Sport Australia's Community Sport Infrastructure Program titled 'Award of Funding under the Community Sport Infrastructure Program'. It found that the award of grant funding was not informed by an appropriate assessment process and that sound advice and the successful applications were not those that had been assessed as the most meritorious in terms of the published program guidelines.[36] Senator Bridget McKenzie, the then Minister for Sport in the Morrison government, was found to have used her ministerial discretion to favour marginal or targeted electorates in the allocation of grants in the lead up to the 2019 Australian federal election.[37]

McKenzie resigned as deputy leader of the Nationals and from her ministerial portfolio on 2 February 2020, but not for the funding scandal. A report by the Secretary of the Department of Prime Minister and Cabinet found that she had breached ministerial standards by not declaring her membership of one of the clubs which had received funding under the program.[38] Further controversy followed with Prime Minister Scott Morrison's rejection of the Auditor-General's initial report and cabinet's decision to hold the Prime Minister's commissioned report confidential.

Bills passed in the Senate 2010–2020

So how did our parliament perform amid the political leadership and scandals and controversies of this past decade? Let's take a look at bills passed in the Senate as an indicator of its performance. If we compare the nine years from 2018–2010 to the prior period 2009–2001 we see an 8 per cent decline in the number of bills passed and a staggering 54 per cent decline in the number of packages of bills passed[39]. Interestingly, the period between 2013–2016 saw the sharpest decline in the number of packages of bills passed by the Senate –from 144 in 2012 to 19 in 2013 and the height of 2,017 total bills passed by the Senate in 2012 decline to its lowest in 2018 (76). The instability of political and party leadership throughout the past decade translated into volatility in the legislative performance of the parliament (see Exhibit 1.5). That period reflected the rivalry between the Labor Party's Rudd/Gillard leadership battle and that of Abbott/Turnbull of the Liberal Party.

Exhibit 1.5: Total number of bills and total number of packages of bills passed by the Senate 2009–2018

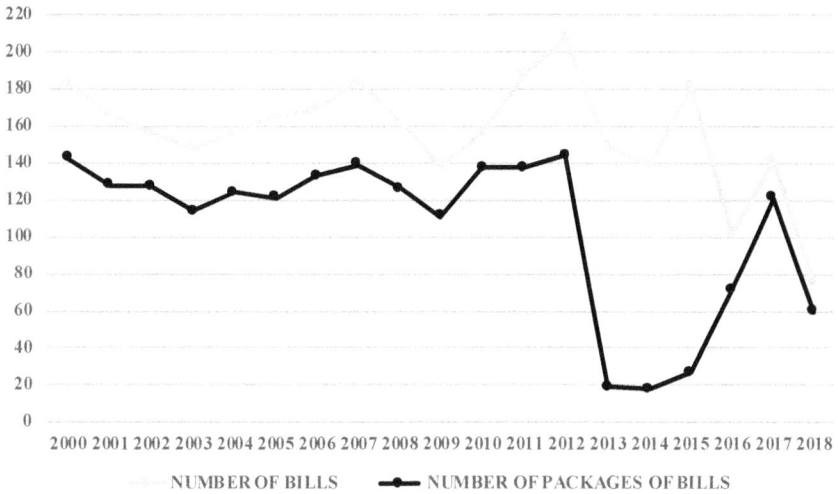

Source: Australian Parliamentary Statistics

Farewell to political giants

Amid the political leadership uncertainty and the scandalous shenanigans throughout that decade, it was also a decade when we farewelled three legendary political leaders: Gough Whitlam (2014), Malcolm Fraser (2015) and Bob Hawke (2019).

For adversaries, they showed us what post political life relationships and friendships can achieve when they united behind some of the biggest issues facing Australia and the world. While this past decade has been characterised by the increasingly oppositional nature of Australian politics, Whitlam and Fraser showed that friendships can be made across party lines. The death of Malcolm Fraser saw senators and members highlight Australia's most high-profile political friendship, that between the two men who had been bitter adversaries in the 1975 dismissal of the Labor government: Gough Whitlam and Malcolm Fraser.

Then Prime Minister Tony Abbott noted in his condolence speech for Mr Fraser:

"Of the 86 condolence speeches for Malcolm Fraser, 25 noted the friendship between the two men, with some urging it to be seen as an example for all

MPs to aspire to." Their friendship was probably helped by their similar stance on broad social issues and that many of the measures Whitlam had initiated were brought to fruition by Fraser, such as the environment. Senator Penny Wong added: "Post Parliament, they were together advocating a 'yes' vote in the 1999 republic referendum; and they sat together as Prime Minister Kevin Rudd made the apology to the stolen generation."

In former Prime Minister Paul Keating's eulogy at Bob Hawke's memorial service, he noted "At the core of it, Bob and I shared one primary idea – that Australia's creativity had been locked down by a stultifying paternal policy regime – the idea that the government knew best and that Australia was best protected and nurtured as a closed economy behind policy barbed wire – a framework that both the Labor Party and the Coalition then heavily subscribed to." This I believe to be just an example of the depth of vision that our former leaders had for Australia, that perhaps seemed absent in the past decade of political leadership.

These colourful leaders have left us with a rich history of classic quotes – here are some of my favourites from each which I believe remain as true today, as they were then when spoken.

''I'd like you to consider: who's the man with his finger on Mr McMahon's autocue? If the gadget fails, if the plug comes out, the whole meeting will come to a standstill. This is going to be Government by teleprompt."

~ Gough Whitlam. Whitlam could not resist having a dig.

"When Government makes opportunities for any of the citizens, it makes them for all the citizens. We are all diminished as citizens when any of us are poor. Poverty is a national waste as well as individual waste. We are all diminished when any of us are denied proper education. The nation is the poorer – a poorer economy, a poorer civilisation, because of this human and national waste."

~ Gough Whitlam, from his 1969 campaign launch.

"It is the first time the burglar has been appointed as caretaker".

~ Gough Whitlam

"There was one sort of trauma in having an election forced. There would have been another sort of trauma in having that Government stay in power for another six of seven months."

~ Malcolm Fraser on replacing the Whitlam government following its dismissal.

"By the year 2020 therefore, we could look to Australia playing an increasing role in east Asian and regional affairs. If our policies were successful, we would see enhanced cooperation between all the countries of the region. We would see the region taking greater charge of its economic future, with an Asian Monetary Fund working to support stability and financial viability. There would be a Political Forum where leaders of east and south-east Asian countries would meet each year to discuss matters of regional concern – an outcome which is long overdue. We could also expect that security between countries in and around the region would be re-enforced because of enhanced respect and increased cooperation amongst member states. We would see a significantly reduced role for the United States in and around the region and a greater reliance on east and south-east Asian relationships."

~ Malcolm Fraser speaking to the Ethnic Community Council of Victoria in 1999.

"The departures from the principles underlining that Liberal Party are substantial and serious. The party has become a party of fear and reaction. It is conservative and not liberal. It has not led positive directions, it has allowed and, some would say, promoted race and religion to be part of today's agenda. I find it unrecognisable as liberal."

~ Bemoaning the direction of the contemporary Liberal Party.
Fraser resigned his life membership of the party he led, in 2009.

"[Malcolm Fraser] went straight from Melbourne Grammar to Oxford. And he would have been a very lonely person, and I think he probably met a lot of black students there who were also probably lonely. I think he formed friendships with them, which established his judgement about the question of colour. That's my theory. I don't know whether it's right or not, but that's what I always respected about Malcolm. He was absolutely, totally impeccable on the question of race and colour."

~ Former Prime Minister Bob Hawke on former Prime Minister Malcolm Fraser

"Any boss who sacks anyone for not turning up today is a bum."

~ Former Prime Minister Bob Hawke, following Australia's victory in the 1983 America's Cup.

The collapse of trust in the government to do the right thing by its people

A politician visited a little remote rural town and asked the locals what the Government could do for them.

"We have two big needs," said the local Mayor. "First, we have a hospital, but no doctor".

Faster than a rat up a drainpipe, the politician whipped out his mobile phone, spoke for a while and then announced, "I've sorted it out. A doctor will arrive here tomorrow.

What is your other need?"

The Mayor replied, we need a mobile tower as there is no coverage in our town"[40].

A longitudinal study by the Scanlon Foundation illustrates the impact of this past political decade on the question of whether federal government can be trusted to do the right thing for the Australian people. The study reveals an unsurprising, but inconvenient truth for our political leadership and major parties[41]. The proportion of respondents indicating they 'almost always' or 'most of the time' trusted government to do the right thing, was at 39 per cent in 2007, the last year of the Howard Government. It rose to 48 per cent in 2009 under the Rudd Government, but from that point it declined sharply. By 2010, amid the of lost confidence in the Rudd Labor Government and entry of the Gillard Government it reached 31 per cent, ultimately reaching its lowest point of 26 per cent in 2012. That represents a 22 per cent decline in trust over just 3 years. The level of trust did rebound slightly, remaining in the range of 29-30 per cent between 2014 and 2019. The proportion selecting 'Almost Never' correspondingly rose during the decade from 13 per cent to 15 per cent (see Exhibit 1.6).

Exhibit 1.6: Q. 'How often do you think the Government in Canberra can be trusted to do the right thing for the Australian people?', 2007–19

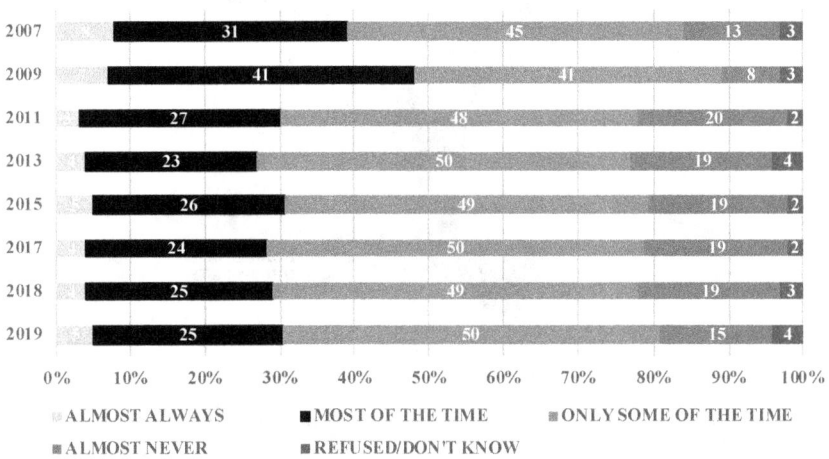

	ALMOST ALWAYS	MOST OF THE TIME	ONLY SOME OF THE TIME	ALMOST NEVER	REFUSED/DON'T KNOW
2007		31	45	13	3
2009		41	41	8	3
2011		27	48	20	2
2013		23	50	19	4
2015		26	49	19	2
2017		24	50	19	2
2018		25	49	19	3
2019		25	50	15	4

Source: The Scanlon Foundation

If the gap between community expectations of our political leaders and their parties, the erosion of trust and political policy volatility performance of our parliament widens in the 2020s, further testing the patience of the electorate may well see the rise of minor parties and independents. Evidence of this manifested in the May 2019 general election, which saw 25 per cent of the electorate not voting Liberal National Party or Labor. The proportion of Australians who had always voted for the same party fell from 69 per cent of voters in 1969 to 39 per cent in 2019. Further, the proportion of Australians who regard themselves as lifetime Liberal National Party voters has fallen from 35 per cent in 1969 to 17 per cent today. Similarly, the proportion considering themselves lifetime Labor voters has fallen from a high of 38 per cent in 1987 to 14 per cent today[42].

Looking ahead to 2030, the Australia 2030 research asked the question 'Who do you trust the most to control your best interests when it comes to the technological and scientific developments over the coming 10 years?' Unsurprisingly, the government ranked the least trusted among a range of other organisational types by one in two people (46 per cent) (see Exhibit 1.7).

Exhibit 1.7: 'Who do you trust the most to control your best interests when it comes to the technological and scientific developments over the coming 10 years? (Least Trusted) (%)

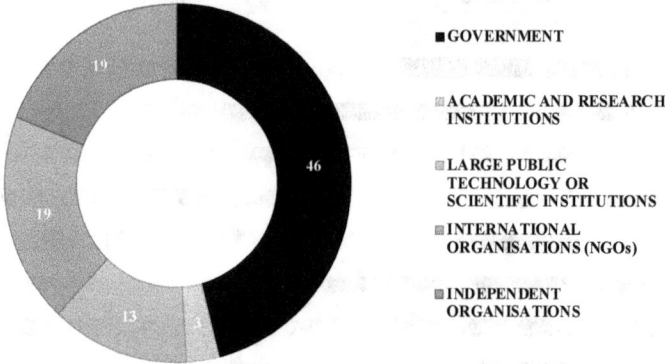

- ■ GOVERNMENT
- ▨ ACADEMIC AND RESEARCH INSTITUTIONS
- ▨ LARGE PUBLIC TECHNOLOGY OR SCIENTIFIC INSTITUTIONS
- ▨ INTERNATIONAL ORGANISATIONS (NGOs)
- ▨ INDEPENDENT ORGANISATIONS

Source: Australia 2030 research Rocky Scopelliti

The research uncovered interesting insights on those who chose the government as the least trusted. Sixty-four per cent of males least trusted the government and older demographic groups were least likely to trust government (Millennials 18–38 years 14 per cent, Generation X 39–53 years 50 per cent and Baby Boomers 54–72 years 36 per cent). Of particular significance, is that 40 per cent of people who classified themselves as leaders including board members, chairman, chief executive officers or business owners least trusted the government to look after their best interests on technological and scientific developments (see Exhibits 1.8, 1.9, 1.10). This is quite interesting given that a study by ANU found that 56 per cent of Australians believe government is only working for "a few big interests"[43]. The Australians most likely to be associated with those 'few big interests" seem to think government isn't doing its job.

Exhibits 1.8, 1.9, 1.10: 'Who do you trust the most to control your best interests when it comes to the technological and scientific developments over the coming 10 years? (Least Trusted) (%) by gender, age and role

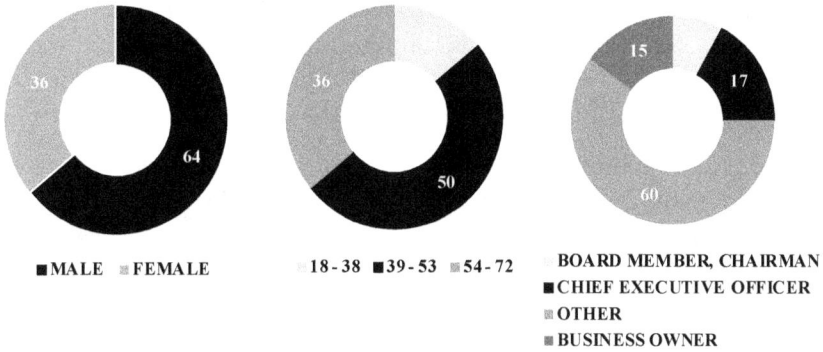

■ MALE   ▨ FEMALE

18 - 38   ■ 39 - 53   ▨ 54 - 72

BOARD MEMBER, CHAIRMAN
■ CHIEF EXECUTIVE OFFICER
▨ OTHER
■ BUSINESS OWNER

Source: Australia 2030 research Rocky Scopelliti
Note: Other includes: Technology & Operations, Finance, Strategy, Sales, Marketing, Product, Legal, Compliance, Risk, Academic, Public Service

The seasonality of Australian politics over the past decade has resulted in Australian professionals lacking confidence that state and federal governments have effective plans in place with industries, the private sector for economic, technological, social and cultural transformation over the coming 10 years. Only 3 per cent feel either extremely, or very confident and only 21 per cent somewhat confident. At the opposite end, 30 per cent reported no confidence at all and 46 per cent not so confident (see Exhibit 1.11).

Exhibit 1.11: 'What confidence level do you have that Australia's Federal and State Governments have effective plans in place with industries and the private sector for economic, technological, social and cultural transformation over the coming 10 years? (%)

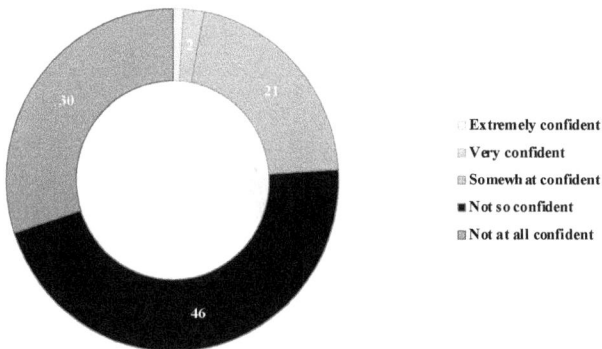

Extremely confident
▨ Very confident
▨ Somewhat confident
■ Not so confident
▨ Not at all confident

Source: Australia 2030 research Rocky Scopelliti

In chapter 8 we will look closely at some of the very significant strategic initiatives and plans aimed at shaping our future. The paucity of such major initiatives and issues with the execution perhaps represent an opportunity lost by Governments to "take the road less travelled" – with paths at key crossroads dictated by political agendas manifesting themselves as these beliefs.

The world of politics

The influence of politics on our attitudes, transcends domestic developments. Throughout the decade there were significant political developments throughout the world. I think a word cloud is a simple and informative way to analyse and communicate much in a single glance, the decade that was the world of politics.

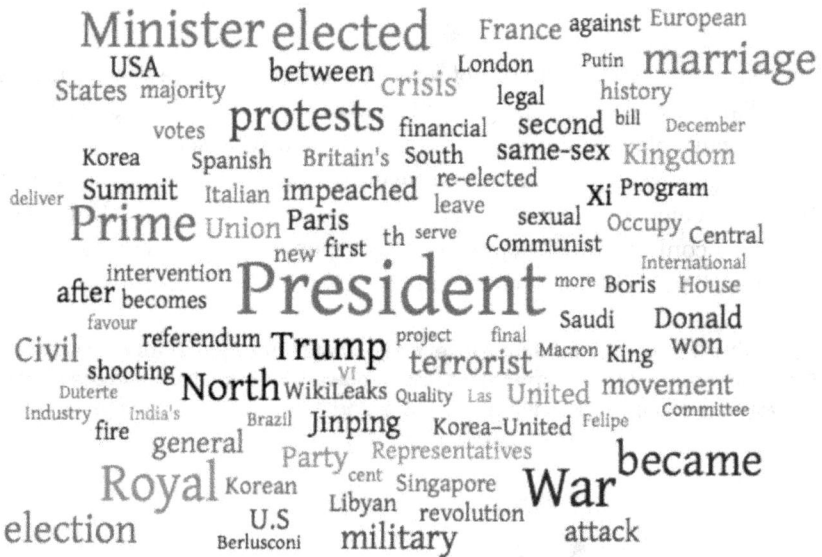

Minister elected France against European USA between London Putin marriage States majority crisis legal history votes protests financial second bill December Korea Spanish Britain's South same-sex Kingdom deliver Summit Italian impeached re-elected Xi Program Prime Union Paris leave sexual Occupy Central new first th serve Communist International intervention President more Boris House after becomes Saudi Donald favour Civil referendum Trump project final Macron King won shooting VI terrorist Duterte North WikiLeaks Quality Las United movement Industry India's Brazil Jinping Korea–United Felipe Committee fire general Party Representatives became Royal Korean cent Singapore War election U.S Libyan revolution attack Berlusconi military

Source: Australia 2030 research Rocky Scopelliti

KEY POINTS

- Australia closed one of the most politically unstable periods in its history. Our nation was divided, disillusioned and disappointed by the political 'self-indulgence' and self-interest demonstrated by politicians from all political sides, including minority parties and independents.
- Despite a majority of Australian voters thinking that the removal of a political party's leader is never justified, major political party's broke world records by changing leaders more often than any other nation. The subsequent animosity between the anointed and displaced leaders, such as Rudd and Gillard and between Abbott and Turnbull became a public spectre serving to destabilise major Australian political parties and their policies.
- Not only did this past decade see a world record of prime ministerial leadership changes, but more political scandals and controversies from both state and federal governments than any other decade in our history. We became sensitised to these leadership spill announcements and scandals and controversies that occupied a large part of the decade's political narrative. That uncertainty has resulted in Australians no longer understanding what each party truly stands for, nor is committed to other than self-interest.
- Australia's parliamentary performance unsurprising bore the brunt of that instability reflected in the significant decline in the number of bills and packages of bills passed that decade compared to the decade prior, particularly during the period between the Labor Party's Rudd/Gillard leadership battle and that of Abbott/Turnbull of the Liberal Party.
- Amid the political leadership uncertainty and the scandalous shenanigans throughout that decade, it was also a decade when we farewelled three legendary political leaders: Gough Whitlam, Malcolm Fraser and Bob Hawke, which showed us the remarkability of leaders to stand above self-interest way beyond their term in office.
- The significant erosion of trust in government from this past political decade to do the right thing for the Australian people highlights the magnitude of the gap between community expectations of our political leaders and their parties. At the last federal election, Labor and Liberal parties saw a significant drop in people who had always voted for the same party. This reflects that Australians, beyond their hopes and dreams, and in the absence of vision, will formulate their own visions rooted in the community.

- Looking ahead to 2030, on the question of 'Who do you trust the most to control your best interests when it comes to the technological and scientific developments over the coming 10 years?', the *Australia 2030 research unsurprisingly reveals that government ranked the least trusted among a range of other organisational types by one in two people. This raises a very interesting question on whether trust in government is irretrievable and that we are now in transition to more contemporary forms of trust.*
- Beyond the lack of trust in government to look after our best interests, the last decade of Australian politics has resulted in Australian professionals lacking any confidence in state and federal governments to have effective plans in place for the coming decade. This brings into question whether government has in fact played a role in 'systems leadership' within the private sector. We will explore this question in detail in other chapters.

We have now crossed two major tipping points when it comes to Australian politics that will set the agenda for the coming decade.

TIPPING POINTS

1. **Electorate support for major political party concentration continues its decline**. Australians have now crossed an inflection point which will see independents and smaller parties elected to provide more political diversity.
2. **Trust and confidence in the government that 'it knows best' when it comes to our future continues its decline.** Australians have crossed the tipping point upon which as a society, we are not prepared to bestow our best interests to political parties when we don't have confidence that they are or willing to act in our best interests.

I'll close this chapter by sharing some of the qualitative quotes from the 382 respondents whose commentary perhaps answers the question posed by Hugh Mackay in 2007 of 'can we improve our political system'?

IN AUSTRALIAN PROFESSIONALS' OWN WORDS

- *"There is a complete lack of visibility of strategic vision and leadership that would give me confidence".*

- *"Federally not at all confident – Show me a 10-year economic, social, or cultural strategy that they have developed in or created in the last six years? There is none that is a feasible long-term plan. State governments seem to be going a much better job at planning the future of their states".*

- *"There does not seem to be a strategic plan as to where we want to be in 10 years' time and therefore what actions are required to make this change happen".*

- *"The level of corruption both seen and unseen"*

- *"I believe, as shown by the current Government they are focused on profit, that's short-term profit and self-gain. I think most lack integrity, as well as are blatantly bureaucratic and out of touch ..."*

- *"Recent history has shown that there is unlikely to be a significant majority in any "house" to be able to significantly develop and implement effective plans. This requires large majorities, and security of tenure. Couple that with the divided accountability between federal and state governments (let alone party politics) and effective delivery of a co-ordinated outcome is unlikely".*

- *"Democracy is in crisis and not adapting to be effective in a social and digital age. The quality of leadership, debate and discourse has lessened over the decades. Policy appears to be driven by polls, by what will win votes, by popularism, rather than by logic and concern for the greater good, the big picture and future implications".*

- *"I work with both the Australian and UK Governments and Australia is being left behind. We organised a meeting with the relevant Australian Government leaders to discuss green finance and one of them thanked us for getting them all in a room together because they had no idea each other existed. Very embarrassing".*

- *"Very low trust in Australian Government, which appears to have regressed in the last decade".*

- *"Our Parliamentary leaders (no disrespect intended) are a monoculture of suburban lawyers, trade union officials and pastoralists – for the most part, they do not reflect Australia in its diversity, nor do they have the knowledge/life experience nor curiosity to understand the wider issues at work both globally and in Australia".*

- *"The past decade has seen little/no leadership from our politicians, who have preferred to spend their time postulating, playing politics and looking after their own interests".*

- *"At a federal level micro-economic reform has almost completely stopped and winning 'political points' has been a bigger focus for the majority of leaders. State Governments across the board seem to be doing much better, though overall with weak federal leadership there isn't a bright future for regulation or policy led business growth in Australia"*

- *"I fear that most past and present leaders are led by good values to make positive change in these areas to only find out that getting votes to remain in parliament requires a disconnect from one or more of these items. Therefore, real and transparent leadership is never attained. We are led by leaders puppeteered by engaging in political action to keep them in power to lead not by political action that is for the greater good of Australia – doing this will ensure their demise of retaining leadership".*

CHAPTER 2

The economy

"Tell him he's dreaming." (The Castle)

'Undeniably, what we are facing today is a very serious situation, but it is something that is temporary. As we deal with it as best we can, we also need to look to the other side when things will recover. When we do get to that other side, all those fundamentals that have made Australia such a successful and prosperous country will still be there. We need to remember that.

To help us get to the other side, though, we need a bridge. Without that bridge, there will be more damage, some of which will be permanent, to the economy and to people's lives.

~ Governor Philip Lowe, Reserve Bank of Australia

This word cloud reflects how we felt about the Australian economy over the past decade and while we were in the eye of the COVID-19 storm and what we think is on the other side.

Source: Australia 2030 research Rocky Scopelliti

In this chapter, we will look at the past decade of economic performance in Australia and then consider the impact of COVID-19 as it relates to the coming decade. Noting of course, that these unprecedented economic conditions are anchored to a world of uncertain economic conditions. COVID-19 in many respects exacerbated the uncertainty of Brexit and is further compounded by uncertainty caused by social instability in the US and Hong Kong just to name a few. One could argue that we are, in fact, seeing positive feedback loops between these various drivers of uncertainty. Is this in fact "exponential uncertainty"?

However, we now know that the decade we are staring into will require a profoundly different set of macro and micro economic settings.

The economy — the worst economic growth figures this century

Australia finished this past decade with an unprecedented cycle:

- Record low interest rates
- Record levels of debt
- Low inflation
- Low unemployment
- Booming exports, and
- Huge government spending.

What made this cycle unprecedented is the fact that all of these have occurred at the same time. We have never seen or experienced this cluster of conditions before. As such, we enter this new decade knowing it will be profoundly different to that of the past decade. Real household income has been stagnant since the Global Financial Crisis (GFC), despite the fact that our national output has been rising. Combating that with record low interest rates, is just not working.

Labour utilisation

The past decade has been showing a rapidly growing gap between unemployment and underemployment. Unemployment (pre-COVID) was relatively static. Underemployment has been growing steadily. There is a very big difference between being unemployed and underemployed. Unemployed means you don't have a job, while underemployment means the job you have is inadequate. Sometimes it is used when talking about people who are working in a lower capacity than one in which they are qualified. However, most often, underemployment is connected to jobs that are lower-paid or for a limited number of hours.

The term is also a measure of labour utilisation. When underemployment is high, the workforce isn't being utilised to its full potential leading to the productivity issue we face today.

The latest labour force figures highlight how great a problem underemployment has become in our economy. The 1 per cent increase in the unemployment rate in April 2020 from 5.2 per cent to 6.2 per cent was the highest ever one-month rise just beating the previous record of 0.9 percentage points set during the 1982 recession. During the same period however, underemployment jumped from 8.8 per cent to 13.7 per cent – a 4.9-per cent increase that dwarfed the previous record one-month increase of 0.8 per cent set during the 1990s recession. Exhibit 2.1 illustrates just how poor the unemployment rate is at portraying the state of the economy over the past decade. The current unemployment rate is not even the highest in the past decade; the current underemployment rate is the highest of all-time by a scarcely believable amount.

Those underemployed who would like to have more hours than they currently work make up approximately 90 per cent of the underemployed. While in April 2020, the number of those who just preferred to work more hours barely changed, the number of those who were underemployed due to economic reasons increased by 400 per cent.

Exhibit 2.1: Underemployment and unemployment rates (%) (2003 – 2019)

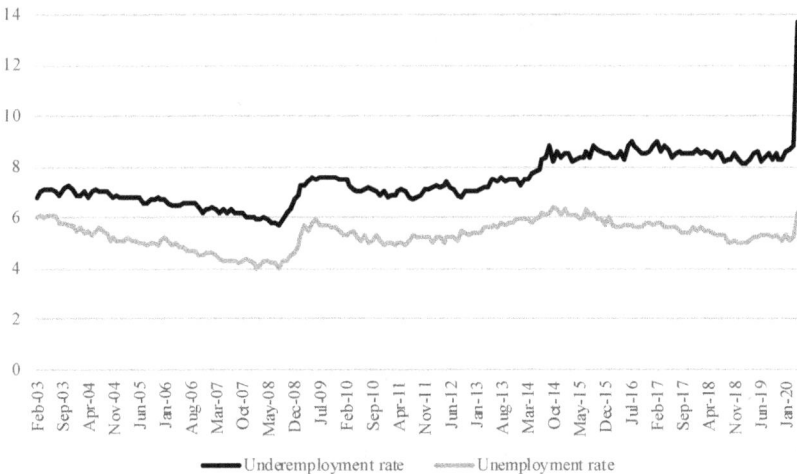

Source: ABS

Australia's first recession in 29 years

In Canberra on 3 June 2020, Treasurer Josh Frydenberg confirmed what we all knew was coming – Australia is in the midst of its first recession in 29 years[44]. The latest GDP figures indicate that the Australian economy declined by 0.3 per cent in the March quarter and expectations that the June quarter would decline a further 8.5 per cent.

Despite Australia comprising just 0.3 per cent of the world's population, its economy was expected to be the fourteenth largest in the world and the fifth largest in the Asian region in 2019. The AUD is also the fifth most traded currency in the world despite the relatively small size of the economy which points to it having a somewhat unique role. Australia's nominal GDP is estimated at almost AUD$2 trillion and accounts for 1.7 per cent of the global economy. In 1999, Australia's total production value was just AU$611, meaning it has more than tripled in two decades.

Australia's economic growth remains lower than at any point in the past two decades and is predicted to continue declining for the coming decade and beyond. Average real gross domestic product (GDP) growth is down to 2.64 per cent in 2019 and forecast to continue declining to 2.3 per cent for 2020–2040 (see Exhibit 2.2)

Exhibit 2.2: Average real GDP Growth (%) (1984 – 2040e)

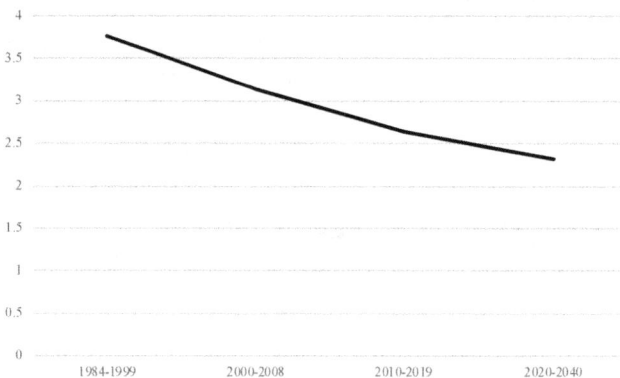

Data Source: OECD

As with declining interest rates that have translated into sentiment based around uncertainty, slower growth predictions can also have the same effect. This results in lower expectations for inflation and real returns that drive decisions about savings, investment and capital allocation which reinforce these same trends. For example, the current discussion about negative interest rates and quantitative easing (QE) may very well be just one manifestation of this unprecedented cycle. However, with a perfect storm of wage stagnation, low growth and stalling productivity, it's time to reset expectations for the decade ahead.

One of the key reasons for Australia's declining economic growth over the past decade, is associated with global growth contraction driven by declining overall global trade activity. For example, over the past decade, GDP growth has declined from historical levels, while trade activity has declined from approximately 61 per cent of global GDP (2008) to approximately 58 per cent in 2019.

Productivity

Declining domestic growth has further created a productivity issue. According to the ABS, growth in output per hour of work, or labour productivity, which had averaged 1.7 per cent a year in Australia since the mid 70s, had halved by 2015 and fell to zero in 2019. Market sector multifactor productivity (MFP) fell 0.4 per cent in 2018–19, the first decline since 2010–11. Market sector gross value added (GVA) grew 1.3 per cent, the slowest output growth recorded for the market sector. By comparison, combined labour and capital inputs grew 1.6 per cent, reflecting capital services growth of 1.8 per cent and hours worked growth of 1.5 per cent. Labour productivity fell 0.2 per cent in 2018–19, the first recorded negative for the sixteen-industry market sector aggregate (since the beginning of the time series in 1994–95)[45] (see Exhibit 2.3).

Exhibit 2.3: Market sector, productivity growth: hours worked basis

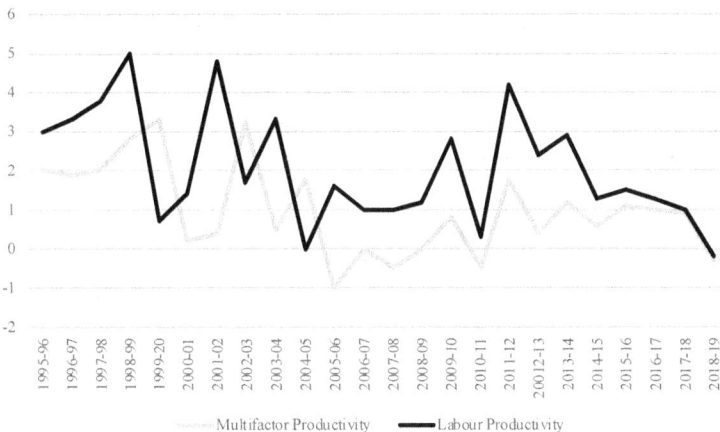

Source: ABS

The US$10 trillion question

Ten years on from the global financial crisis, the global economy remains locked in a cycle of low or flat productivity growth despite the injection of more than US$10 trillion by central banks. According to the World Economic Forum, most economies are still locked in a cycle of low or flat productivity growth. However, what they reported in their 2019 Global Competitiveness report was that economies that have channelled investments into human capital, improving institutions, innovation capability and business dynamism will be best placed to revive productivity and withstand a global slowdown[46].

Globally, we've reached an economic tipping point with subdued growth, rising inequalities and accelerating climate change provide the context for a backlash against capitalism, globalisation, technology, and elites. There is gridlock in the international governance system and escalating trade and geopolitical tensions are fuelling uncertainty. This holds back investment and increases the risk of supply shocks: disruptions to global supply chains, sudden price spikes or interruptions in the availability of key resources.

Then came the virus and bedrock assumptions about the Australian and world economy changed overnight.

The unfolding economic crisis caused by the COVID-19 pandemic is most often compared to the 'Great Depression of the 1930s' or the recession of 1991. But the run up to the Great Depression, was characterised by a massive boom where a surging economy in the United States made way for excess and mass consumerisation which has gestated over many years, compared to the relatively quick run up with COVID-19. Australia may have concluded the past decade with an unusual cluster of economic conditions, but for the decade ahead, we face a completely new rule book – one that is yet to be written.

According to the OECD, growth prospects remain highly uncertain[47]. In March 2020, it lowered its global growth outlook by 0.5 per cent from its November 2019 forecast. This reflects the adverse impact on confidence, financial markets, key industry sectors and disruption to supply chains particularly those that are strongly interconnected to China, such as Japan, Korea and Australia. For Australia, the revised forecasts indicate a decline of 0.5 per cent in GDP growth in 2020 and a further decline of 0.3 per cent forecast for 2021 to 2.6 per cent. However, the OECD forecasts do suggest Australia will recover much faster than other comparable economies such as Canada, Japan, Korea, United Kingdom. But these are very difficult to predict. For example, many analysts predicted the immediate impact of iron ore exports that didn't materialise. No-one could really have predicted that Brazil – Australia's major competitor for iron ore, would be that heavily impacted in its iron ore production. Effectively Vale – the world's largest producer has completely shut down production. In fact, major Australian producers are reporting to be on track for their best quarter ever. Another major factor is that China has reopened, and the government is mandating not just a return to normal production, but actually catching up the lost production. (see Exhibit 2.4).

Exhibit 2.4: OECD Interim Economic Outlook Forecasts, 2 March 2020 – Real GDP Growth (%)

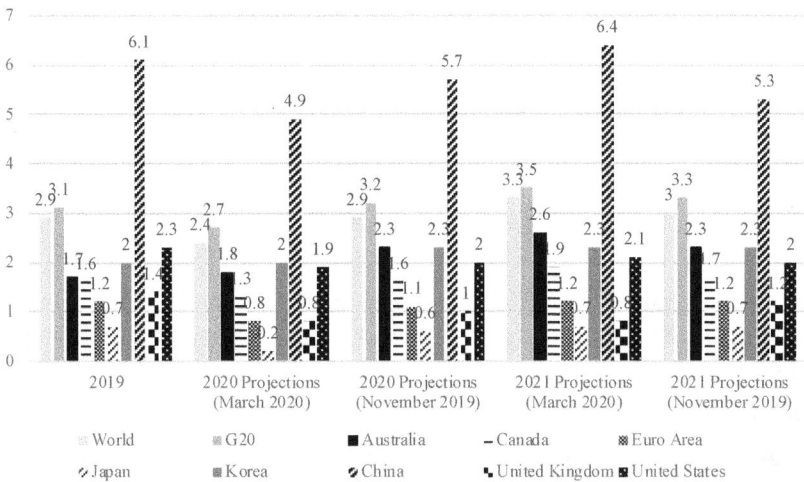

Data Source: OECD

With highly targeted and effective policies and assuming COVID-19 subsides, 2021 may see a recovery with GDP growth forecasted around 3.25 per cent. For example, the Reserve Bank cut interest rates to a record low 0.25 per cent, provided a $105 billion lending facility to banks to support small business. Quantitative easing measures that included government three-year bonds to ensure the base cost of borrowing for corporate and government would stay around 0.25 per cent[48].

That same week saw unprecedented announcements by federal and state governments with stimulus and rescue packages totalling AU$89.5 billion. These measures represent approximately 10 per cent of the Australian economy. Then on 30 March, the Morrison government announced the $130 billion JobKeeper Payment to help keep Australians in jobs as they dealt with the significant economic impact from COVID-19. This brought the federal government's total support for the economy to $320 billion across the forward estimates, representing 16.4 per cent of annual GDP[49] (see Exhibit 2.5).

Exhibit 2.5: Stimulus and rescue packages announced by Australian Federal Government during March 2020 (A$320m)

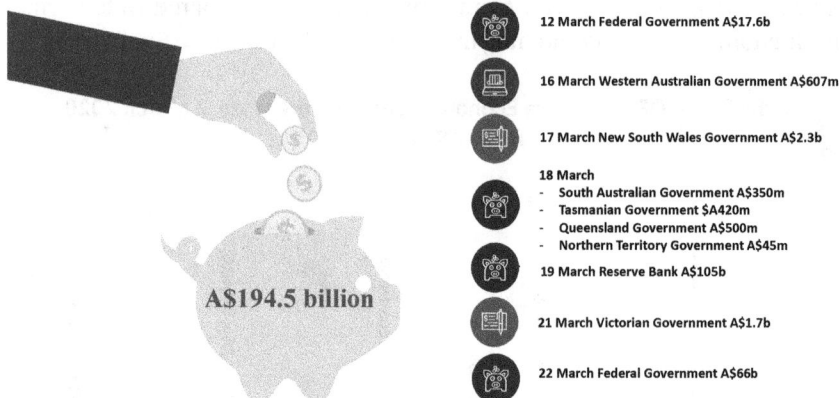

12 March Federal Government A$17.6b

16 March Western Australian Government A$607m

17 March New South Wales Government A$2.3b

18 March
- South Australian Government A$350m
- Tasmanian Government $A420m
- Queensland Government A$500m
- Northern Territory Government A$45m

19 March Reserve Bank A$105b

21 March Victorian Government A$1.7b

22 March Federal Government A$66b

A$194.5 billion

The Morrison government's stimulus, welfare and support packages for the economic shock that the Treasurer Josh Frydenberg described as 'deeper, wider, longer' include:

- Supporting business investment
- Providing cashflow assistance to help small and medium sized business and not for profit to stay in business and keep their employees in jobs
- Access to unsecured loans, wages subsidies
- Targeted support for the most severely affected sectors, regions and communities
- Household welfare stimulus payments that will benefit the wider economy.

These packages were aimed to directly support 6.5 million individuals and 3.5 million businesses. Treasurer Josh Frydenberg said "Australia is approaching the economic challenge from COVID-19 from a position of strength with IMF and the OECD both forecasting Australia to grow faster than comparable countries including the UK, Canada, Japan, Germany and France"[50].

The federal government introduced a wage subsidy program to support employees and businesses. The JobKeeper Payment was designed to help businesses affected by COVID-19 to cover the costs of their employees' wages, so that more employees can retain their job and continue to earn an income.

The strategy was based on keeping Australians in work and businesses in business so that it would lay the foundations for a stronger economic recovery once the COVID-19 crisis passed.

At a state government level, packages included more support for small businesses, households and communities through:

- Boosting health care
- Provide business support and jobs
- Invest in major new infrastructure
- New tourism infrastructure
- Target industries under duress such as hospitality, tourism.

According to a Roy Morgan study in mid-March 2020 of 1,000 + businesses[51], more than 60 per cent of Australian businesses were affected by COVID-19 – a massive 45 per cent increase from the prior month. At an industry level, the study found that several industries have been hit especially hard by the COVID-19. For example, businesses in manufacturing (78%), wholesale trade (74%), recreation & personal (83%), information media & telecommunications (75%), property & business services (74%) and transport, postal & warehousing (72%) report being affected by COVID-19 to some degree. The state-by-state data suggests that South Australia has been hit hardest, with 68 per cent of businesses reporting consequences, followed by NSW and Victoria (see Exhibit 2.6).

Exhibit 2.6: Businesses affected by COVID-19. March 2020 compared to February 2020 (%)

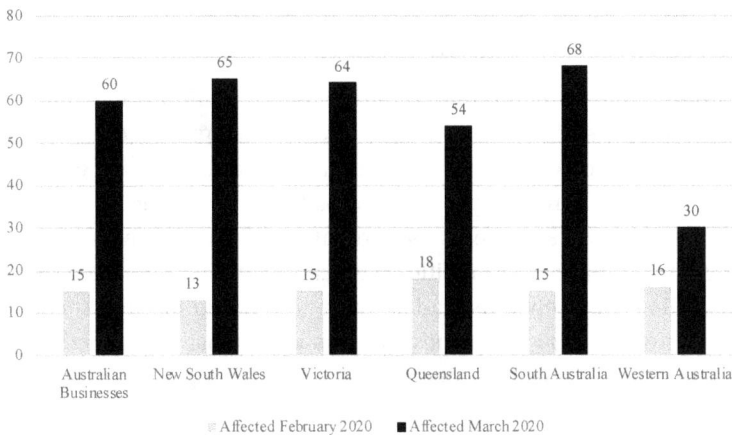

Source: Roy Morgan Special Snap SMS Poll of Australian businesses in Feb. 2020, n=1,170, March 2020, n=1,148. Base: Australian businesses

The reality and the speed at which Australian professionals recognised COVID-19 was not just a health issue, but indeed an economic one, was evident by those respondents in the Australia 2030 research who indicated that low economic growth would affect the world and Australia over the next 10 years. During the months of January and February 2020, between 13–21 per cent of respondents ranked 'low economic growth' as the number one issue which skyrocketed to 95 per cent of those respondents in March 2020 (see Exhibit 2.7).

Exhibit 2.7: Q. In your opinion, which of the following issues will affect the world, and Australia over the next 10 years? (%)

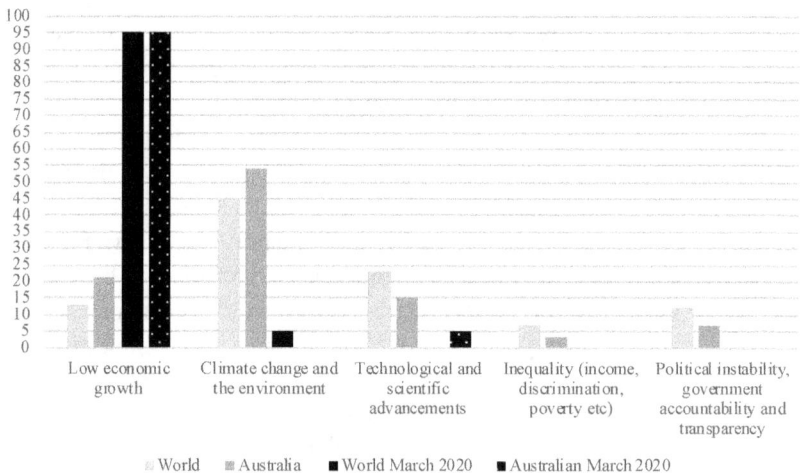

Source: Australia 2030 research Rocky Scopelliti

Interestingly, across the total study period, gender difference was observed. Seventy-two per cent of males ranked low economic growth the number one issue, compared to 26 per cent of females. Older demographic groups were more concerned than younger age groups (Millennials 18–38 years 15 per cent, Generation X 39–53 years 49 percent and Baby Boomers 54–72 years 36 per cent). Of particular significance again, is that 37 per cent of people who classified themselves as leaders including board members, chairman, chief executive officers or business owners were most concerned with 'low economic growth' impacting Australia over the coming decade (see Exhibits 2.8, 2.9, 2.10)

Exhibits 2.8, 2.9, 2.10: 'In your opinion, which of the following issues will affect Australia over the next 10 years? Respondents who ranked 'Low Economic Growth' number one over three months by gender, age and role

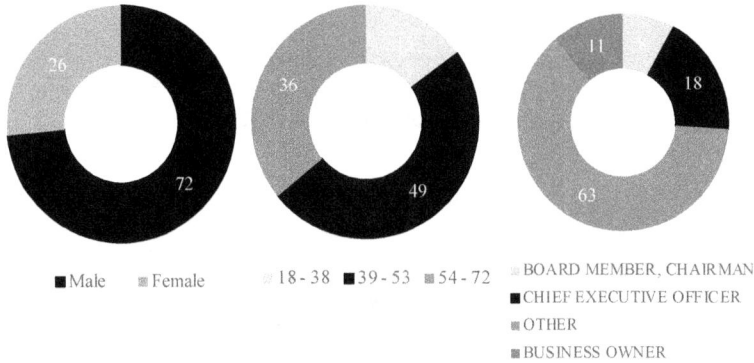

■ Male   ▨ Female   ▨ 18 - 38   ■ 39 - 53   ■ 54 - 72   ▨ BOARD MEMBER, CHAIRMAN
■ CHIEF EXECUTIVE OFFICER
▨ OTHER
■ BUSINESS OWNER

Source: Australia 2030 research Rocky Scopelliti
Note: Other includes: Technology & Operations, Finance, Strategy, Sales, Marketing, Product, Legal, Compliance, Risk, Academic, Public Service

For those who ranked 'low economic growth' as the most significant issue to affect Australia over the coming 10 years, 10 per cent believe that their organisation would not likely be around in the next 10 years. They expressed concerns across a range of issues:

- 74 per cent do not have confidence that Australia's federal and state governments have effective plans in place with industries and the private sector for economic, technological, social and cultural transformation over the next 10 years
- 72 per cent are concerned that Australia isn't investing enough in technological, scientific and skills development compared to other countries for the coming 10 years
- 45 per cent believe that values-led policy change including integrity, honesty, transparency, humility and accountability will be the most important leadership qualities for world leaders over the coming 10 years
- Just 9 per cent trust government the most to control your best interests when it comes to the technological and scientific developments over the coming 10 years. The most trusted were reported to be 'Academic Research Institutions' (34 per cent), followed by 'Independent Associations or Groups' (27 per cent) and 'Large Public technology or Scientific Institutions' (23 per cent)
- Financial institutions ranked fifth (7 per cent) as predicted to be the most ethical, honest and transparent over the coming 10 years. International

organisations (eg NGOs) ranked first (28 per cent) followed by employers second (27 per cent), academic institutions third (24 per cent), government fourth (10 per cent) and big tech companies last at 5 per cent (see Exhibit 2.11).

Exhibits 2.11: 'In your opinion, which of the following issues will affect Australia over the next 10 years? (Rank 1 = Most, 5 = Least)? (%) Respondents who ranked 'Low Economic Growth' 1

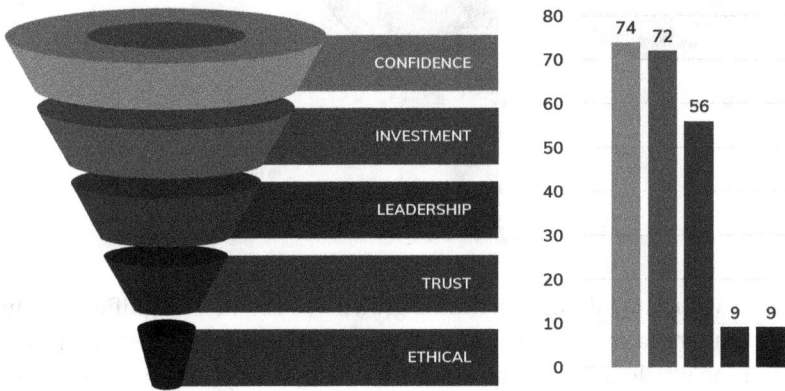

CONFIDENCE

INVESTMENT

LEADERSHIP

TRUST

ETHICAL

80 — 74 72

70

60 — 56

50

40

30

20

10 — 9 9

0

Source: Australia 2030 research Rocky Scopelliti

The economic eye of the storm

In the five weeks since restrictions were imposed to 18 April 2020, almost one million people lost their jobs with an unemployment rate around 12 per cent. The overall wages bill and number of jobs dropped 8.2 per cent and 7.5 per cent respectively during that time[52]. Fears of the unemployment rate shooting up to 20 per cent and impacting 2.3 million people unless these restrictions were lifted within six months of them being imposed were being expressed by the Australian Business Council[53].

According to the Business Council of Australia, its modelling indicates that the economic effects of COVID-19 are likely to be more severe the longer the shutdowns continued. Its modelling shows a 10.1 per cent plunge in Australia's gross domestic product for 2020 if restrictions were unwound from May 2020. A delay until July 2020 could cause a 14.3 per cent plunge in GDP while waiting until October 2020 could cause a 20.7 per cent plummet in Australia's economic output. Treasury estimates the economy will lose $4 billion for every week COVID-19 restrictions remain in place (see Exhibit 2.12).

Exhibit 2.12: Recovery profiles of different timelines. GDP Impacts, deviation from 'no virus' scenario, CY2020 (monthly) and 2021 (quarterly).

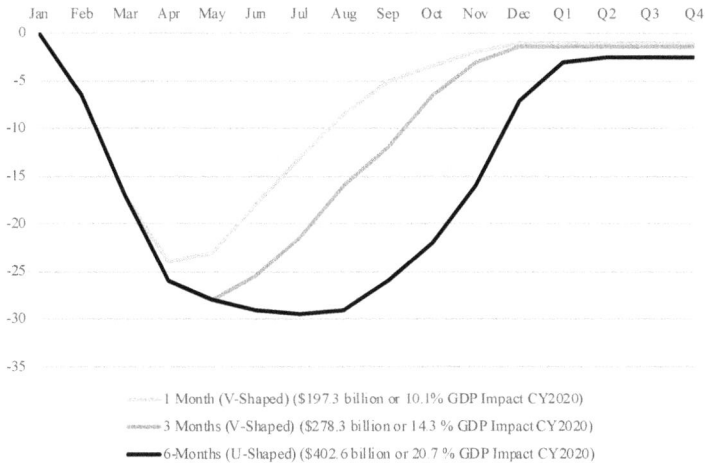

- - - - 1 Month (V-Shaped) ($197.3 billion or 10.1% GDP Impact CY2020)
━━━━ 3 Months (V-Shaped) ($278.3 billion or 14.3 % GDP Impact CY2020)
━━━━ 6-Months (U-Shaped) ($402.6 billion or 20.7 % GDP Impact CY2020)

Data Source: ABC

To appreciate the significance and leadership of the Morrison government's response to COVID-19, its direct fiscal stimulus response was the highest of all advanced economies. As at May 2020, Australia ranked number 1 among 46 other countries, that combined, had announced more than $13 trillion in stimulus measures in response to COVID-19 with approximately four times GDP direct fiscal response, relative to the median across advanced economies (see Exhibit 2.13).

Exhibit 2.13: Direct fiscal stimulus advanced economies (May 2020) (% GDP)

COUNTRY	DIRECT FISCAL MEASURES	LOANS AND GUARANTEES
AUSTRALIA	10.6	1.9
CANADA	5.2	3.3
FRANCE	0.7	13.9
GERMANY	4.4	29.6
ITALY	1.2	32.4
KOREA	1.5	6.4
SPAIN	1.2	9.8
UK	3.1	15.7
USA	6.9	4.2
DENMARK	2.7	3.0
FINLAND	1.7	5.1
NETHERLANDS	2.7	-
NORWAY	2.6	3.4
SINGAPORE	7.0	4.1
SWEDEN	2.2	5.0
MEDIAN	2.7	5.1

Source: BIS Oxford Economics

What's on the other side anyway?

"We can't keep Australia under the doona."

~ Scott Morrison, Australian Prime Minister[54]

This will be the decade of debt. Economists are forecasting the federal government will have $850 billion of bonds on issue and net debt of more than 30 per cent of the $2.2 trillion GDP by 2023[55]. That implies that it will take decades to achieve a budget surplus, but economists suggest that this is a medium-term problem and that government shouldn't respond to that debt and deficit forecast too quickly. We should remember that government is issuing all this debt at real negative interest rates, in effect being paid to borrow.

The term 'exponential' has now found its rightful place within the very fabric of how we've come to understand the pace of change, and importantly, the speed upon which we need to adapt. While I'll cover this more compressively in chapter 8 as it applies to technological and scientific development, economics also follows the same logic. COVID-19 has challenged the linear models and assumptions underpinning the design of systems including health, education, supply chain, workplace, scientific, technological, transport, logistics, financial, social welfare, regulatory and many other systems and practices. The test of that was our capacity to respond to the challenges, with the required speed as the COVID-19 crisis unfolded.

These systems and practices were designed, developed and evolved on linear based thinking of incrementalism that has permeated the 1st, 2nd, and 3rd Industrial Revolutions. But COVID-19 reminds us that we have now transitioned into the 4th Industrial Revolution that according to the World Economic Forum, will impact every nation, industry, organisation and individual[56]. The speed, scale and impact of the 4th Industrial Revolution is what makes it vastly different from those of the past. This conundrum was so well articulated by Jack Welch, Chemical Engineer, writer and former Chairman and CEO of General Electric as he described the consequences for organisations of not sustaining the required pace of change": *'If the rate of change on the outside, exceeds the rate of change on the inside, the end is near.'* In other words, economic performance is not necessarily a function of its access to scare resources, but indeed its capacity to adapt and respond to rapidly changing conditions in the external environment.

The key point here is that using yesterday's logic to prepare economies and indeed polices may not set us on the right trajectory to emerge in a good state the other side. We need new thinking as history has no precedent for what has occurred through COVID-19. Assuming this or other disasters or crisis won't occur again is simply not a feasible position. The one key learning we must now build upon from COVID-19, is that many of our systems did not cope adequately at the beginning of the crisis and in the eye of it. In the absence of change, they will remain inadequate on the other side. We need new thinking on the capacity of all of our major systems in a world characterised by 'exponential impact' and 'exponential uncertainty' – we are no longer in a linear world.

'Pandemics aren't linear – they are exponential

~ *William Hynes, Head of the New Approaches to Economic Challenges Unit, OECD*

In my last book *"Youthquake 4.0 – A Whole generation and the New Industrial Revolution"*, I described the impact of this exponential world which is challenging our notions of value creation in-so-far as it is being dematerialised, decentralised and disintermediated through technological and scientific advances. The resulting effect is that traditional economic theory based on supply and access to resources is becoming decoupled from the way we think about development, growth and exponential models. I described this as *'a new economic physics'.*

According to Rana Foroohar, a global business columnist and an associate editor at the Financial Times, based in New York and CNN's global economic analyst, economists need to abandon their 'linear' based comfort zones[57]. Foroohar argues that as the pandemic plays out, we will require "more creative and complex thinking than we see in most mainstream economics today". Foroohar's point being that we need to evolve our black and white traditional economic 'comfort zones' thinking of systems that rely on customs, history, and time-honoured cultural values and beliefs that guide decisions of production and distribution, to one that is 'variegated'.

According to William Hynes, the head of an OECD unit called New Approaches to Economic Challenges (NAEC), "Pandemics aren't linear – they are exponential. When things get knocked off track, they don't always come back to a steady state. We're talking about complex systems." [58]

NAEC's research on integrative economics puts people at the centre of economic policy and utilises the insights and methods of a range of disciplines. These are integral to painting a realistic picture of how the economic system is shaped and helps shape the larger systems that it is part of. It argues that traditional economics does not provide the tools we need to analyse the many, often irrational-seeming behaviours that are generated by the uncountable interactions of billions of people, firms and institutions locally or globally, in small groups or as nations, at timescales ranging from nanoseconds to millennia. It observes that these human systems are complex and prone to cascading failure. As such, the more policymakers attempt to optimise these systems, the more unstable they become. These systems – financial markets, production chains and the economy – are built for short-term efficiency not long-term resilience. So, while the efficiency of these systems may be disputable, many parts in the system result in the concentration of vital industrial capacities.

As Australia prepares to come out from under the 'doona' there seems to be a general presumption that companies, markets and indeed the world, will eventually reset to normal, albeit perhaps a "new normal" and that efficiency rather than resiliency is where equilibrium is to be found. While building resilience in our economic systems is harder than promoting productivity alone, it may well be what defines the shape we are in when we get to the other side. For Australia, taking the 'road less travelled' through the limitations of assumptions developed on a linear view of recovery, may not see us arrive on the other side as well as we could have, had we taken the 'road not taken'.

So, what's on the other side? More resilient integrative systems designed on new economic physics that can adopt to exponential events.

KEY POINTS

- Australia finished the decade with an unprecedented economic cycle with record low interest rates, record levels of debt, low inflation, low unemployment, booming exports and huge government spending. What made this cycle unprecedented, is the fact that all of these have occurred at the same time.
- Declining domestic growth has created a productivity issue. Growth in output per hour of work, or labour productivity, which had averaged 1.7 per cent a year in Australia since the mid 70s, had halved by 2015 and fell to zero in 2019. Market sector multifactor productivity (MFP) fell 0.4 per cent in 2018–19, the first decline since 2010–11. Market sector gross value added (GVA) grew 1.3 per cent, the slowest output growth recorded for the market sector. By comparison, combined labour and capital inputs grew 1.6 per cent, reflecting capital services growth of 1.8 per cent and hours

worked growth of 1.5 per cent. Labour productivity fell 0.2 per cent in 2018–19, the first recorded negative for the sixteen-industry market sector aggregate (since the beginning of the time series in 1994–95).

- With highly targeted and effective policies and assuming COVID-19 subsides, 2021 may see a recovery with GDP growth forecasted around 3.25 per cent. For example, the Reserve Bank cut interest rates to a record low 0.25 per cent and provided a $105 billion lending facility to banks to support small business. Quantitative easing measures that included government three-year bonds to ensure the base cost of borrowing for corporate and the government would stay around 0.25 percent.

- The reality and the speed at which Australian professionals recognised COVID-19 was not just a health issue, but indeed an economic one, was evident by those respondents in the Australia 2030 research who indicated that low economic growth would affect the world and Australia over the next 10 years. During the months of January and February 2020, between 13–21 per cent of respondents ranked 'low economic growth' as the number one issue, which skyrocketed to 95 per cent of those respondents in March 2020.

- In the eye of the storm, the economic effects of COVID-19 were likely to be more severe the longer the shutdowns continued. Modelling shows a 10.1 per cent plunge in Australia's gross domestic product for 2020 if restrictions were unwound from May 2020. A delay until July 2020 could cause a 14.3 per cent plunge in GDP while waiting until October 2020 could cause a 20.7 per cent plummet in Australia's economic output. Treasury estimates the economy will lose $4 billion for every week COVID-19 restrictions remain in place.

- This will be the decade of debt. Economists are forecasting the federal government will have $850 billion of bonds on issue and net debt of more than 30 per cent of the $2.2 trillion GDP by 2023. That implies that it will take decades to achieve a budget surplus, but economists suggest that this is a medium-term problem and that government shouldn't respond to that debt and deficit forecast too quickly. We should remember that the government is issuing all this debt at real negative interest rates, in effect being paid to borrow.

- The term 'exponential' has now found its rightful place within the very fabric of how we've come to understand the pace of change, and importantly, the speed at which we need to adapt. COVID-19 has challenged our linear designed and centred health, education, supply chain, workplace, scientific, technological, transport, logistics, financial, social welfare, regulatory and many other systems and practices. The test of that was our limited capacity and speed at which to adapt and the many challenges that presented themselves globally.

- Pandemics, like many other global crises, aren't linear – they are exponential. When things get knocked off track, they don't always come back to a steady state. Traditional economics does not provide the tools we need to analyse the many, often irrational-seeming behaviours that are generated by the uncountable interactions of billions of people, firms and institutions locally or globally, in small groups or as nations, at timescales ranging from nanoseconds to millennia. These human systems are complex and prone to cascading failure. As such, the more policymakers attempt to optimise these systems, the more unstable they become. These systems – financial markets, production chains and the economy – are built for short-term efficiency not long-term resilience.
- Integrative economics puts people at the centre of economic policy and utilises the insights and methods of a range of disciplines. These are integral to painting a realistic picture of how the economic system is shaped and helps shape the larger systems that it is part of.
- As Australia prepares to come out from under the 'doona' there seems to be a general presumption that companies, markets and indeed the world, will eventually reset to normal and that efficiency rather than resiliency is where equilibrium is to be found. While building resilience in our economic systems is harder than promoting productivity alone, it may well be what defines the shape we are in when we get to the other side. For Australia, taking the 'road less travelled' through the limitations of assumptions developed on a linear view of recovery, may not see us arrive on the other side as well as we could have, had we taken the 'road not taken'.

TIPPING POINTS

1. **National debt will be a stable guest in our economy** and we need to get used to its presence. It's not going anywhere fast and awaits us on 'the other side' for the decade ahead.
2. **Transforming our systems** will allow a more integrative set of economic settings that allow the economy to respond in a more agile manner, as and when the next global crisis occurs.

The world of economics

The influence of economics on our attitudes is significantly influenced by the world we live in. Throughout the decade there were significant economic developments throughout the world. This word cloud thematically represents the decade that was the world of economics.

economies growing **decade** development national tariffs central
Asian Century **global crisis**system **United**
growth globalization deficit
politicians foreign stability
China's President inequality sanctions
rates policy Europe WTO interest Trump's future **tariffs**
China-led resources
public imports power emerging rules-based Brexit FTA **economic**
inflation government **China** trading Asia oil leadership
political free American IMF **banks** shape
workers multilateral reaction administration serious intellectual
regional World **trade** influence market encouraged governments
EU investment agreements **policies**
Brexit subsidies banking markets order GDP Euro **investment**
unilateralist America's century **production** cryptocurrency
leverage **Bank** property **States** fakenews
countries **Trump** regime First Chinese income **international**

Source: Australia 2030 research Rocky Scopelliti

I'll close this chapter by sharing some of the qualitative quotes from respondents who ranked low economic growth as Australia's number one issue and provided commentary on the question of confidence in government and its plans to 2030.

IN AUSTRALIAN PROFESSIONALS' OWN WORDS

- *"I believe that we are moving towards a massive recession in Australia, party due in fact we acted on the GFC when we really didn't need to, so this limits the Government, partly in due to everything runs in cycles it always has and we in some ways need a recession to clean out the weak businesses so the strong can thrive and provide our countries next cycle of growth, also we are the most indebted country of private debt to GDP in the world so something is going to give soon. My problem is Government these days don't know how to leave the economy alone, it will work itself out. The Government should help people in need in this down turn but not try and kick the can down the road again, because at some point it's going to happen and if you keep kicking the can it just makes it bigger when we do hit the recession".*

- *"No proven track record in keeping pace with the current rate of change".*

- *"The Government is now approximately 30 per cent of the economy and ineffective in social change unless forced. Businesses and communities need to rely on each other more than Government".*

- *"For the late 30 years we have failed to reform energy, housing, health, education, tax. And under-invested in innovation".*

- *"Recent history has shown that there is unlikely to be a significant majority in any "house" to be able to significantly develop and implement effective plans. this requires large majorities, and security of Tenure. Couple that with the divided accountability between Federal and State Governments (let alone party politics) and effective delivery of a co-ordinated outcome is unlikely".*

- *"Government seems to be pandering to those with the most money. If it isn't in their interests for transformation, then it may not be a priority".*

- *"Our Australian politics is still caught up in petty matters; they are not taking leadership position in global issues".*

- *"Failure to invest in new technologies, failure to invest in infrastructure that would support new ways of working and living. Continued failure to invest in education at all levels. No genuine leadership".*

- *"Too much focus to get macro growth and core economic stimulus over micro stimulation".*

- *"Still not clear on the strategic plan for our economy – maybe because 'Scomo' has not provided one to the public"?*

- *"We have the recent management of our emergency of fires. Poor understanding, poor messaging and care for the public. Disconnected".*

- *"Governments have talked big, spent unwisely and do not have the delivery experience to work with private enterprises. Large projects get taken for a costly ride, whilst smaller, much more worthy projects are put through the same processes and therefore fail to get through all the hurdles".*

CHAPTER 3

The environment

"That's bullshit.
You just bullshitted NASA."
(The Dish)

"If you ignore the science when you build a bridge,
the bridge falls down."

~Professor Ross Garnaut, economist

This word cloud thematically analyses and represents how we felt about the environment over the past decade.

Fitzsimmons leadership national
carbon tax CO2levels industry policy
inconvenient concern
carbon bushfires experts Scott scientists increasing
sentiment
fire weather increase Australia's politicisation
denial
Minister polluters truth electorate service risk major Prime
leaders resource sector experts global carbon
nationalsecurity Throughout Australian warming listen threats
renewable bushfire denial opportunity floods million fires
nation political threat climate study leadership Indigenous
fakenews former public nature solar crisis Commissioner
risks change bullshit services Morrison world decade backlash
among science
Australia Party tax government emergency
inconvenient truth effects action significant
drought levels support forces security events
science political ignorance renewable
political ignorance Australians catastrophic
energy frequency self combusting cow manure natural
leader Garnaut

Source: Australia 2030 research Rocky Scopelliti

In this chapter, we will explore the impact that the politicisation of climate change has had over the past decade. The significance of that correlation between policy, or its failings and public sentiment is critical for Australians in order to chart a new course for this coming decade. Sadly, we closed the decade with what some described as an apocalyptic bushfire season. Climate change has become Australia's inconvenient truth and this truth is dislocated from our history season coming towards the tail of what is by many measure, in many regions the worst drought in recorded history. Aboriginal and Torres Strait Islander people understand there is an intricate and inextricable interconnection between the physical, chemical and biological sciences and the social sciences more widely – long before industrial revolutions of the modern world understood this. This is particularly so because of the deep and timeless relationship between Country and Aboriginal and Torres Strait Islander identities, languages, cultures and spiritualties.

Climate change leadership

What does good leadership look like? Let's begin by reflecting on one of the most remarkable (among many) leaders who stepped forward, leaned in, and lead us through the New South Wales bushfires of the summer of 2019/20, the worst hit state. None other than Craig Fitzsimmons – Commissioner of the Rural Fire Service (RFS) in NSW.

Commissioner Fitzsimmons became the face of the crisis, and a household name. He was the person we trusted and turned to for warnings and key developments, in a tell-it-like-it-is, informative, respectful, compassionate and empathetic communication manner. Nicknamed the "nation's father" last year, Commissioner Fitzsimmons became the comforting face of a devastating 2019-2020 bushfire season, from which NSW is still reeling.

Sadly, Commissioner Fitzsimmons had to tell three families their loved ones would not be returning when firefighters Andrew O'Dwyer, Geoff Keaton and Samuel McPaul died this year. As a nation we felt his emotion when it came time to share the heartbreaking news with the nation. The images of pinning a medal onto Harvey Keaton, whose father died in the bushfires, and his embrace of other children and family members touched us all. He understood their grief, as his own father, George, had died in a fire in a national park in 2000.

His frankness reflected the political sentiment of the nation, when he revealed publicly that he was "disappointed and frustrated" at Prime Minister Scott

Morrison for not communicating the decision to deploy 3,000 army reservists in the support effort to communities. This honesty, so often on display by Commissioner Fitzsimmons, was respectfully acknowledged by the prime minister as he told reporters in Nowra that he would work to have better communications with Commissioner Fitzsimmons[59].

"You are the only person that I and probably many of the people of Australia have any trust in at the moment."

~ Tweet from a member of the public on RFS Twitter to Commissioner Fitzsimmons

According to reports, Professor Ariadne Vromen, a political sociology expert from the University of Sydney, commented that the leadership comparison between Prime Minister Scott Morrison and Commissioner Fitzsimmons was especially highlighted by the Australian media. "Fitzsimmons has been constantly present unlike the Prime Minister. He's portrayed as not only the professional leader, but also someone who can demonstrate empathy, who can show his own emotion in a way people find relatable." [60]

The report noted that Commissioner Fitzsimmons has also emphasised the role of climate change with the bushfires. "There is that kind of honest and truth-telling that is resonating with people, particularly when it contrasts with the obfuscation from the federal Government" said Professor Ariadne Vromen.

For the decade ahead, Australia has an opportunity to step forward, lean in, and show leadership on our fragile environment – some would say a second opportunity – a second chance to take the "road less travelled" we previously shied away from. Throughout this Chapter I will build on this theme and question of who are our policy makers are listening to, if not our experts?

Australia's climate change policies
– our inconvenient political truth

Over the past 14 years, The Lowy Institute has surveyed Australians about their views on global warming and climate change[61] (see Exhibit 3.1). This longitudinal study shows a steady decline between 2006 – 2012, in public concern that global warming is a serious and pressing and that we should start taking steps to address it – even if it involves costs. They offer the possible explanation in the political infighting and leadership churn in Australia during the Rudd/Gillard era, as well as a fierce campaign against a 'carbon tax' by the opposition. During that same period the study found a corresponding increase in the public's uncertainty that climate change was a pressing issue, or that it was an actual problem.

Exhibits 3.1: Australia and Climate Change (%)

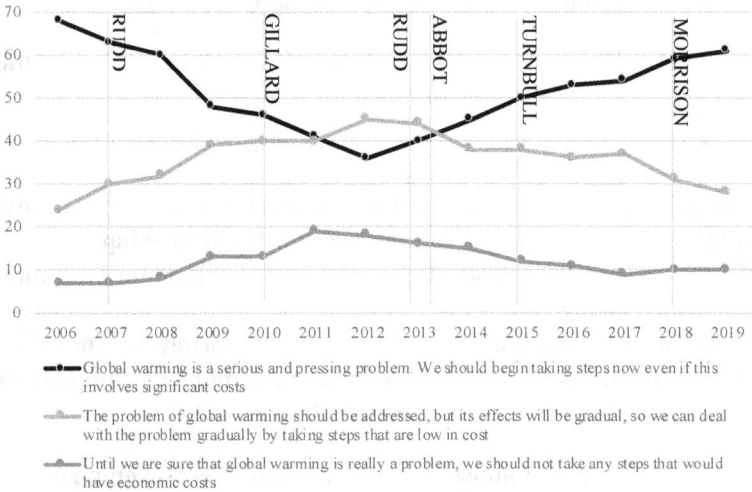

Source: The Lowy Institute

The Lowy Institute study highlights the significant change in Australians' attitudes since 2012 with concern returning to the same level as 2007. The study found that in 2019, 61 per cent of Australians reported global warming as 'a serious and pressing problem' about which 'we should begin taking steps now even if this involves significant costs'. It found that the issue of climate change and global warming continues to split Australians along generational lines. A clear majority of 76 per cent of young Australians aged 18–44 agree with this statement, compared to just half (49 per cent) of their elders.

Throughout that period of time, Australia recorded its driest year on record with the drought lasting to 2009. As leader of the then opposition Labor Party, Kevin Rudd declared that 'climate change is the greatest moral challenge of our generation', which then became a major campaign platform for the Labor Party. After winning the 2007 federal election, the first official act of the newly elected Rudd government was to ratify the Kyoto Protocol. The following year the Labor Party committed to a Carbon Pollution Reduction Scheme (CPRS) to commence in July 2010 and to reduce Australia's greenhouse gas emissions by at least 5 per cent from 2000 levels by 2020. After a failed attempt to pass CPRS legislation through the senate, Prime Minister Rudd announced in 2010 that the legislation would-be put-on hold until 2012.

In a 2010 leadership challenge resulting in Julia Gillard displacing Kevin Rudd as Prime Minister and leader of the Labor Party, Gillard promised there would be no carbon tax under her government. Shortly after, the federal election in September that year produced a hung parliament. Gillard formed a minority government with key independents, but importantly, signing an agreement with the Australian Greens that included climate changes. The following year (2011), with support from the Greens and key Independents, the Australian parliament passed the Clean Energy Act 2011, which laid the foundations for a carbon price mechanism. Leader of the Opposition at that time Tony Abbott, commenced a campaign against the legislation committing to repeal it should the Liberal National Party win office.

In 2013, the Abbott government was elected to office and in 2014, the carbon price mechanism was repealed with the support of the crossbench. Globally, Australia was recognised as being the first nation to reverse policy action on climate change. In 2015, after a leadership spill by the Liberal Party, Malcolm Turnbull displaced Tony Abbott becoming prime minister and leader of the Liberal National Party. Three months later, Australia agreed to adopt the Paris Agreement in which all nations submit and review national emissions reduction pledges with the aim to keep global temperature increases below 2 degrees Celsius. Australia became one of 175 countries to sign the Paris Agreement in 2016.

What did our scientists and experts advise?

For over a decade, politicians were warned by our scientists about the 2020 consequences of climate change. Twelve years ago, economist Professor Ross Garnaut, one of Australia's most distinguished and well-known economists, led an independent study of the impacts of climate change on the Australian economy. In the final report, he made a prediction that has devastatingly come true. The 2008 Garnaut review commissioned by Kevin Rudd found that the "fire seasons will start earlier, end slightly later, and generally be more intense" and that "this effect increases over time, but should generally be observable by 2020"[62] (see Exhibit 3.2).

Exhibit: 3.2 Extract from 'The Garnaut Climate Change Review – Final Report'

Bushfires

Recent projections of fire weather (Lucas et al. 2007) suggest that fire seasons will start earlier, end slightly later, and generally be more intense. This effect increases over time, but should be directly observable by 2020.

Table 5.4 shows projections of the percentage increase in the number of days with very high and extreme fire weather.[6]

Table 5.4   Projected per cent increases in the number of days with very high and extreme fire weather for selected years

	Approximate year		
	2013	2034	2067
Very high	+2–13	+10–30	+20–100
Extreme	+5–25	+15–65	+100–300

Note: This study was based on scenarios producing 0.4°C, 1.0°C and 2.9°C temperature increases, which equate to the years in this table under a no-mitigation case.

Source: Lucas et al. (2007).

Source: The Garnaut Climate Change Review – Final Report page 118

In an interview in January 2020, Garnaut was asked for his reaction to the then bushfires impacting Australia.

"It's one of sadness, that I was ineffective. Having been given the opportunity to talk to Australians on this issue, that I was ineffective in persuading Australians that it was in our national interest to play a positive role in a global effort to mitigate the effects of climate change," said Garnaut[63].

On Saturday 7 February 2009, 173 people lost their lives and more than 2,000 houses were destroyed in bushfires in the Australian state of Victoria. That year, Australia's national science research agency, the CSIRO, and the Bushfire and Natural Hazards Cooperative Research Centre (CRC) released a study prepared for the 2009 Senate Inquiry into Bushfires in Australia[64], which warned that the kind of rare weather event fuelling the fires of that year—a particular low-pressure system colliding with a particular high-pressure system—would be up to four times more likely under forecast climate change-related warming.

A month before that tragic event, the CRC issued a press release including the following statements on 12 January 2009:

"Australia's Chief Fire Officers now believe that our current knowledge and practices on bushfire management will not meet the expected needs of the community in coming decades".

"Climate change and drought are altering the nature, ferocity and duration of bushfires and an ageing and declining volunteer population are challenging the way fire agencies are going to be able to manage these events. These issues are being further compounded by the expanding rural urban fringe and the desire for people to retire to these semi-rural or rural areas. These demographic changes mean there will be increasing numbers of people living in these higher risk zones that are less capable of dealing with the fire risk".

"When we consider recent events, it is not hard to imagine a repeat of Black Friday. The Victoria fires of 2006 were the longest campaign fires in recorded history. And the Canberra fires of 2003 showed the devastation that can be caused by this type of runaway bushfire, how under certain conditions even urban regions can support considerable fire spread entering urban zones within them."

The 2009 royal commission into the causes of the state of Victoria's devastating Black Saturday bushfires, alluded to the fact that bushfire risks were only likely to increase. A 2017 Climate Council report found that climate change was increasing the severity and intensity of bushfires.

At a global level the World Economic Forum's Global Risks Report 2020[65] identified that the top five risks to the world were environmental (see Exhibit 3.3). Severe threats to our climate account for all of the Global Risks Report's top long-term risks, with "economic confrontations" and "domestic political polarization" recognised as significant short-term risks in 2020. The report warns that geopolitical turbulence and the retreat from multilateralism threatens everyone's ability to tackle shared, critical global risks. The report alerted leaders that without urgent attention to repairing societal divisions and driving sustainable economic growth, leaders cannot systemically address threats like climate or biodiversity crises.

Exhibit: 3.3 Major risks to the world

	Low	High
Risk Likelihood	• Weapons of mass destruction	• Climate action failure
	• Information infrastructure breakdown	• Biodiversity loss
		• Natural disasters
	• Infectious diseases	• Water crisis
	• Food crisis	• Extreme weather
	• Financial failure	• Cyber attacks
	• Fiscal crisis	• Human-made environmental disasters
	• Unemployment	• Global governance failure
	• Critical infrastructure failure	• Interstate conflict
	• State collapse	• Involuntary migration
	• Terrorist attacks	• Social instability
	• Energy price shock	• Data fraud or theft
	• Unmanageable inflation	• Asset bubbles
	• Failure of urban planning	
	• Deflation	
	• Adverse technological advances	
	• National governance failure	
	Low	High

Risk Impact

Data Source: World Economic Forum

Climate change – our inconvenient political truth

Throughout this decade, atmospheric carbon dioxide was reaching levels unprecedented in modern times, with record temperatures to match. On 9 May 2013, global CO2 levels reached 400 parts per million for the first time in human history, and by 2016, CO2 levels were staying firmly above this threshold. As a result, the world felt an uptick in warming; 2015, 2016, 2017, 2018, and 2019 were the five hottest years on record since 1880. For Australia, the past decade shows that the annual mean temperatures were the highest on record at 1.52°C above the average (see Exhibit 3.4)

Exhibit 3.4: Australian mean temperature anomaly (1910–2019) (°C)

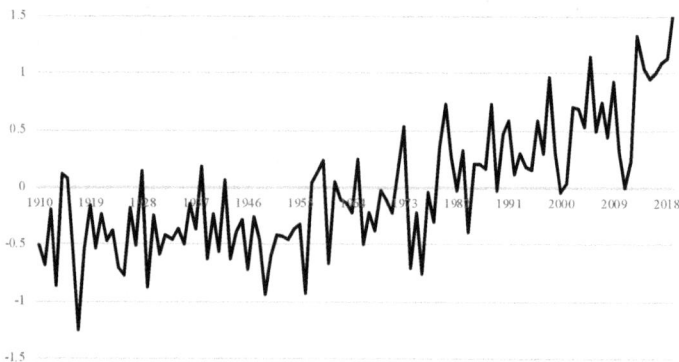

Data Source: Australian Bureau of Meteorology

In a series of major reports, the world's scientists forcefully called attention to earth's altered climate, the risks it poses, and the need to respond. In 2014, the Intergovernmental Panel on Climate Change (IPCC) released its fifth assessment of climate change's reality and consequences[66]. A year later, the world's nations negotiated the Paris Agreement, the global climate accord that aims to keep warming below 2 °C—which world leaders and scientists consider a dangerous threshold. In October 2018, the IPCC published another grim report[67] that outlined the huge costs of warming even 1.5 °C by 2100—which is likely the minimum the planet will face. In the face of such huge challenges, record-breaking climate protests have swept the world, many led by youth activists.

'The country is ground zero for the climate catastrophe.'[68]

~Richard Flanagan, author

How do we now feel about climate change?

Like many Australians and indeed, people the world over, we were saddened by the bushfires that bookmarked our closure on this past decade and the commencement of the new. We were inspired by the courage of volunteers, emergency services and the Australian spirit that gave comfort and reassurance and unified the nation in ways that no other event did during that decade. We were humbled by the support from countries around the world including firefighters, aircraft, care for the flora and fauna and much more. An indicator of the significance of this event to Australians can be illustrated by the extent to which we were all engaged in other major events throughout the decade. According to news reports and analysis by Google searches, Australians searched the bushfires more than any other news event in the past decade[69] (see Exhibit 3.5). Nationwide, we were very engaged.

Exhibit: 3.5 Most Trending News Searches 2010–2020 (%)

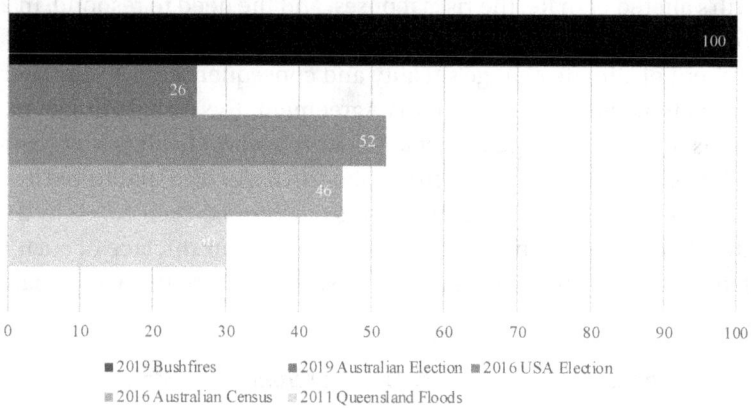

| | | | | | | | | | | 100 |
| 0 | 10 | 20 | 30 | 40 | 50 | 60 | 70 | 80 | 90 | 100 |

Bar values shown: 100, 26, 52, 46

■ 2019 Bushfires   ■ 2019 Australian Election   ■ 2016 USA Election

■ 2016 Australian Census   ■ 2011 Queensland Floods

Data Sources: Google, ABC

According to Google search data, interest in our Prime Minister Scott Morrison became a trending topic domestically and internationally during the fires – for all the wrong reasons. Was this a backlash of his choice to holiday before the peak of the bushfire catastrophe? Perhaps a one-off public reaction? Or was it a reflection of the discord on climate change that had accumulated over the course of the decade between the public, who now look to our scientists and experts, and successive major political parties and their inconvenient truth?

The Australia 2030 research conducted over the period between January–March 2020 revealed that Australians see 'Climate Change and the Environment' as by far the single biggest issues that the world and Australia face over the coming decade – 2.5 times greater that 'Low Economic Growth' or 'Advancements in Technology and Science' (see Exhibit 3.6)

Exhibit: 3.6 Q In your opinion, which of the following issues will affect the world/ Australia over the next 10 years? (%)

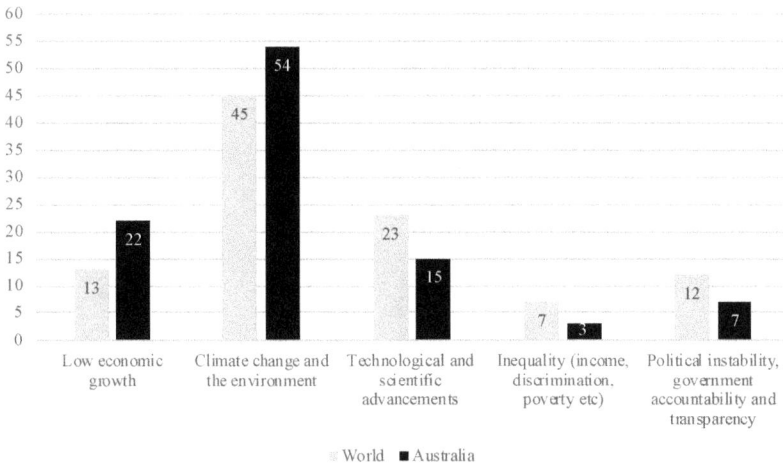

	Low economic growth	Climate change and the environment	Technological and scientific advancements	Inequality (income, discrimination, poverty etc)	Political instability, government accountability and transparency
World	13	45	23	7	12
Australia	22	54	15	3	7

World ■ Australia

Source: Australia 2030 research Rocky Scopelliti

Among Australian professionals, there was very little generational, gender or professional difference on this issue in relation to the world or Australia. This suggests that when people think about the impact of climate change, we have a unified view. The past decade saw Australian policy become detached from the broader global position, yet we remain a 'follower' where public sentiment is clearly providing the opportunity for Australia to lead.

Beyond comparing the significance of climate change to those issues in this study, other studies reveal a similar finding. Over the past 15 years, the Lowy Institute has sought Australians' views on a range of possible threats to 'the vital interests of Australia in the next ten years', including terrorism, the nuclear threat from unfriendly countries, immigration, fake news and China's growing power. For the first time in the history of the Lowy Institute Poll, climate change topped the list of threats to Australia's vital interests in 2019, followed by cyberattacks, international terrorism and North Korea's nuclear program (see Exhibit 3.7).

The study reveals that the past five years have seen a dramatic reversal in Australian attitudes regarding climate change and its impact. This reversal has returned to points not seen since 2008. In 2019, 61 per cent of Australians say, 'global warming is a serious and pressing problem', about which 'we should begin taking steps now even if this involves significant costs'.

Exhibit: 3.7 List of possible threats to the vital interests of Australia in the next ten years

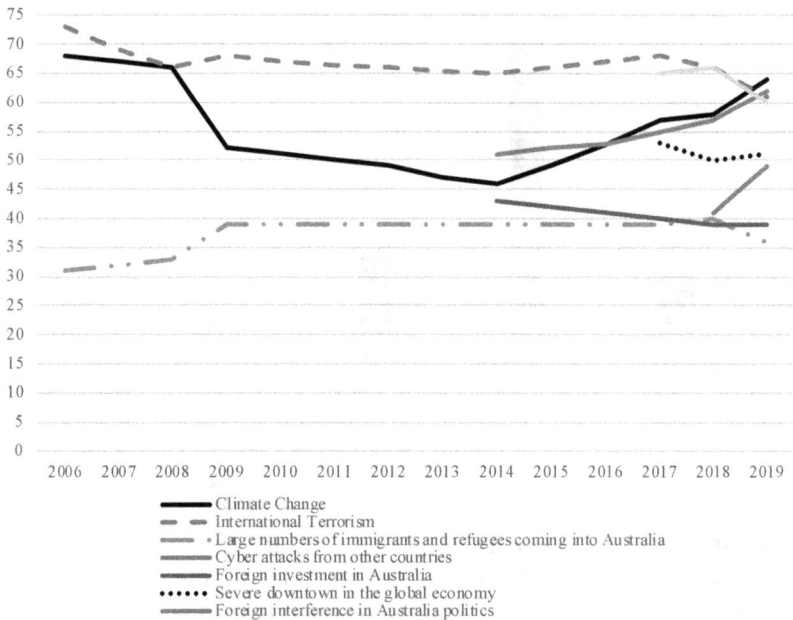

Climate Change
International Terrorism
• • Large numbers of immigrants and refugees coming into Australia
Cyber attacks from other countries
Foreign investment in Australia
••••• Severe downtown in the global economy
Foreign interference in Australia politics

Source: Lowy Institute Poll 2019

The devastation across Australia caused by the bushfires referred to previously has had a far-reaching impact for Australians, our flora, fauna and the broader environment. It was reported that[70]:

- An estimated 11.3 million Australians or 57 per cent of the Australian adult population were physically affected by smoke
- An estimated 10.6 million Australians, or over half the adult population, suffered anxiety or worry for the safety of themselves, their close family or their friends
- More than 12.6 million hectares of forest destroyed
- More than 1 billion animals killed
- More than 434 million tonnes of carbon dioxide (CO_2) emitted into the atmosphere.

Who can forget the images of Toni Doherty who took off her shirt to rescue a koala from burning? Sadly, the koala, named Ellenborough Lewis after the area in which it was found, died due to the severity of its burns. That image and many more of the bravery of firefighters, and the public, together with the unprecedented scale and severity of the bushfire crisis, has raised rightful questions in Australian society about the inconvenient truth and the politicisation and subsequent inaction on climate change, and why successive governments in the past decade did not listen to our experts.

The scientific community has long reported, argued and predicted that for Australia, climate change means a significant increase in the occurrence and size of bushfires. Twenty-three former fire and emergency leaders say they tried for months to warn Prime Minister Scott Morrison that Australia needed more water bombers to tackle bigger, faster and hotter bushfires[71].

For Australia, climate change will lead to an increase in the number of high-risk fire days over a year, an increasing amount of dry fuels and solids, and a declining window to engage in fuel reduction burning exercises. The 2019 drought and the summer of 2019/20 have now drawn our attention to the fact that climate change is no longer an issue to be addressed in the future, but rather it has arrived.

The sharp increase over the past five years in Australian attitudes regarding climate change and its impact, has now surpassed those of the threat from Korea's nuclear program, international terrorism and state-sponsored cyber-attacks. Yet despite that, successive governments did not reflect community sentiment in their key statements about the strategic environment and threats to Australian national security. In the 2013 National Security Statement and the 2016 Defence White Paper, climate change merits only passing mention. With 3,000 army personal deployed to support our RFS and communities during the 2019/20 bushfires, there is no doubt a closer look at the risk's climate change poses to the nation is warranted.

The security implications of climate change now mean long-term preparedness and planning to manage its effects for Australia's defence forces. Deployment of our armed forces has been critical to both the floods and bushfires over the past year. If anything, the bushfires have reinforced already growing public concern about climate politics and climate security. When we reflect on this public concern, we should recognise that for Australians, the threat is more direct and immediate. A changed climate guarantees large-scale disasters, rising sea levels, reduced rainfall, and devastating effects on our way of life. By way of contrast, China, long held out as a laggard on climate change, has since 2017 explicitly identified climate change as a significant national security issue.

Australian policymakers have the opportunity to listen to our experts, to the electorate and to unify and acknowledge the threat posed by climate change. If climate change is prioritised among our national security threats, it will encourage progressive bipartisan policy making that addresses the threats themselves and the rights and needs of Australians. If anything positive can come from the catastrophic bushfires, it will be to end the toxic politics of climate change in Australia over the past decade. Now is the time to listen to the electorate, the science, and our remarkable service men and women. For policy makers of all persuasions, as you stand at your crossroads, the decade ahead will see Australians of all walks of life, take the 'road not taken'. Come join us, but it cannot entail the same decision-making paradigms leadership model of the past. We need the same systems leadership we saw throughout the COVID-19 response as these increasingly random and extreme climatic events will require us to be far more agile in our responses. I will respectfully close this chapter with the letter from Emergency Leaders for Climate Action[72], who in April 2019, warned major political parties that Australia was unprepared for worsening natural disasters from climate change and that governments are putting lives at risk. These are the same leaders and experts, like Shane Fitzsimmons, who we looked to for leadership in our time of need throughout the 2019/20 bushfire season.

That letter was written in 2019, before the last federal election in May of that year and calls on the next prime minister of Australia to meet former emergency service leaders "who will outline, unconstrained by their former employers, how climate change risks are rapidly escalating". That next prime minister became Scott Morrison, and despite their efforts to warn him, by November 2019 (six months later), they accepted that they would not get that meeting.

In 2007, Hugh Mackay challenged us to the question of 'are we serious about climate change?' Thirteen year later, many would say yes, we are, but our political leaders and their parties aren't. When amid our greatest national catastrophe, Australia's Acting Prime Minster believed that 'self-combusting piles of manure' is a cause of bushfires, it's time for Australians to listen to one another, our experts and the facts. Acting Prime Minister Michael McCormack attracted worldwide attention when addressing a press conference, he summarised the link between climate change and the unfolding bushfire catastrophe as follows.

*'Climate change is not the only factor that has caused these fires. There have been dry lightning strikes, **there has been self-combusting piles of manure,** there has been a lot of arsonists out there causing fire.*[73]'

~ Michael McCormack – Acting Prime Minister
(while Prime Minister Scott Morrison was holidaying in Hawaii)

Joint Statement from Emergency Leaders for Climate Action

Australia Unprepared for Worsening Extreme Weather

We, the undersigned, who are former senior Australian fire and emergency service leaders, have observed how Australia is experiencing increasingly catastrophic extreme weather events that are putting lives, properties and livelihoods at greater risk and overwhelming our emergency services. Climate change, driven mainly by the burning of coal, oil and gas, is worsening these extreme weather events, including hot days, heatwaves, heavy rainfall, coastal flooding and catastrophic bushfire weather. Australia has just experienced a summer of record-breaking heat, prolonged heatwaves, and devastating fires and floods—there should be no doubt in anyone's mind: climate change is dangerous, and it is affecting all of us now.

Facts You Need to Know

- Bushfire seasons are lasting longer and longer.
- The number of days of Very High to Catastrophic bushfire danger each year are increasing across much of Australia and are projected to get even worse.
- Opportunities to carry out hazard reduction burns are decreasing because warmer, drier winters mean prescribed fires can often be too hard to control – so fuel loads will increase.
- Higher temperatures mean that forests and grasslands are drier, ignite more easily and burn more readily, meaning fires are harder to control.
- 'Dry' lightning storms are increasing in frequency, sparking many remote bushfires that are difficult to reach and control.
- Fire seasons across Australia and in the northern hemisphere used to be staggered – allowing exchange of vital equipment such as aerial water bombers, trucks and firefighters. The increasing overlap of fire seasons between states and territories and with the USA and Canada will limit our ability to help each other during major emergencies.
- A warmer atmosphere holds more moisture, increasing the risk of heavier downpours and flooding events—like that which recently affected Townsville.

- Current federal government climate policy has resulted in greenhouse gas pollution increasing over the last four years, putting Australian lives at risk. Communities, emergency services and health services across Australia need to be adequately resourced to cope with increasing natural disaster risk.

Tackling climate change effectively requires rapidly and deeply reducing greenhouse gas pollution here in Australia and around the world. We have the solutions at our disposal, we just need the political will to get on with the job.

We call on the Prime Minister to:

- Meet with a delegation of former emergency services leaders who will outline, unconstrained by their former employers, how climate change risks are rapidly escalating.
- Commit to a parliamentary inquiry into whether Australian emergency services are adequately resourced and equipped to cope with increasing natural disaster risks due to climate change.
- Recognise that strategic national firefighting assets like large firefighting aircraft are prohibitively expensive for states and territories, are leased from the northern hemisphere, and that increased overlap of fire seasons is restricting access to this equipment during times of need. A cost-benefit analysis of current arrangements and their effectiveness, and how Australia's strategic aerial firefighting needs can be best met and funded, needs to be initiated in consultation with the National Aerial Firefighting Centre.
- Ensure continued funding for stakeholder-driven research into how we can respond to, mitigate, and increase resilience to bushfires, natural hazards and escalating climate change risks.

We call on all State and Territory Governments to:

- Provide increased resources to enable forestry, national parks, urban and rural fire services to increase environmentally sensitive fuel reduction and fire mitigation programs.
- Focus on climate change adaptation and mitigation programs while taking strong action to significantly reduce state / territory emissions.
- Cease cutting the budgets and resources of forestry, national parks, urban and rural fire services, both directly and through instruments such as "efficiency dividends", so that the services can increase operational capacity to deal with our "new normal" of catastrophic weather risks.

This joint statement is signed by:

Mary Barry
Former CEO, Victorian State
Emergency Service

Neil Bibby AFSM
Former Chief Executive Officer,
Country Fire Authority Victoria, and
former Deputy Chief Officer,
Melbourne Metropolitan Fire Brigade

Tony Blanks AFSM Former Fire
Unit Manager, Tasmania National
Parks, and former Fire Manager,
Forestry Tasmania

Mike Brown AM, AFSM Former
Chief Fire Officer, Tasmania Fire
Service

Naomi Brown
Former CEO, Australasian Fire &
Emergency Service Authorities
Council

Bob Conroy
Former Fire Manager, NSW National
Parks and Wildlife Service

**Major General Peter Dunn AO
(Ret)** Former Commissioner, ACT
Emergency Services Authority

John Gledhill AFSM Former Chief
Fire Officer, Tasmania Fire Service

Dr Jeff Godfredson AFSM Former
Chief Fire Officer, Melbourne
Metropolitan Fire Brigade

Dr Wayne Gregson APM Former
Commissioner, WA Dept of Fire &
Emergency Services

Craig Hynes AFSM Former Chief
Operations Officer, WA Fire and
Emergency Services Authority

Lee Johnson AFSM Former
Commissioner Qld Fire & Emergency
Services. Director: Bushfire &
Natural Hazards Cooperative
Research Centre

Murray Kear AFSM Former
Commissioner, NSW State
Emergency Service

Phil Koperberg AO, AFSM BEM
Former NSW Minister for the
Environment, former Commissioner
NSW Rural Fire Service

Craig Lapsley PSM
Former Emergency Management
Commissioner and Fire Services
Commissioner, Victoria, former
Deputy Chief Officer, Country Fire
Authority

Victoria Andrew Lawson AFSM
Former Deputy Chief Officer, SA
Country Fire Service

Grant Lupton AFSM Former Chief
Fire Officer, South Australian
Metropolitan Fire Service

Greg Mullins AO, AFSM Former
Commissioner Fire & Rescue NSW.
Climate Councillor

Frank Pagano AFSM, ESM
Former Executive Director,
Emergency Management Queensland,
and former Deputy Commissioner,
Queensland Fire & Rescue Service

Steve Rothwell AFSM Former
Director and Chief Fire Officer, NT
Fire & Emergency Services

Stephen Sutton
Former Chief Fire Control Officer,
Bushfires NT

Ken Thompson AFSM Former
Deputy Commissioner, Fire & Rescue
NSW

Ewan Waller AFSM Former Chief
Fire Officer, Forest Fire
Management, Victoria

Source: ELCA (Emergency Leaders for Climate Action) (2019) Emergency Leaders: Australia Unprepared for Worsening Extremes. Accessed at https://www.climatecouncil.org.au/emergency-leaders-climate-action/.

KEY POINTS

- Climate change has become Australia's political inconvenient truth and this truth is dislocated from our history. Aboriginal and Torres Strait Islander people understand there is an intricate and inextricable interconnection between the physical, chemical and biological sciences and the social sciences more widely – long before industrial revolutions of the modern world understood this. This is particularly so because of the deep and timeless relationship between Country and Aboriginal and Torres Strait Islander identities, languages, cultures and spiritualties.
- The extent of the politicisation of climate change by major Australian political parties throughout the past decade became evident when Australia

became the first nation to reverse policy action on climate change. This was in spite of politicians being warned by our own scientists before the start of the decade about the 2020 consequences of climate change.

- Among Australian professionals, the Australia 2030 research, like many other studies, found there was very little generational, gender or professional difference on the issue of the importance of climate change and the need to act now. This suggests that when people think about the impact of climate change, we have a unified view. The past decade saw Australian policy become detached from the broader global position, yet we remain a 'follower' where public sentiment is clearly providing the opportunity for Australia to lead.

- The national security implications of climate change now mean long-term preparedness and planning to manage its effects for Australia's defence forces. Deployment of our armed forces was critical to both the floods and bushfires over the past year. If anything, the bushfires have reinforced already growing public concern about climate politics and climate security. When we reflect on this public concern, we should recognise that for Australians, the threat is more direct and immediate. A changed climate guarantees large-scale disasters, rising sea levels, reduced rainfall, and devastating effects on our way of life. Towns deserted and farms abandoned are examples of that hardship.

- The opportunity for Australian policymakers is to unify and acknowledge the threat posed by climate change, prioritise it among our national security threats, and encourage progressive bipartisan policy making that address the threats themselves and the rights and needs of Australians.

- If anything positive can come from the catastrophic bushfires, it will be to end the toxic politics of climate change in Australia over the past decade. Now is the time to listen to the electorate, the science, and our remarkable service men and women.

We have now crossed two major tipping points when it comes to climate change that will set the agenda for the coming decade.

1. **The national security implications of climate change** now require long-term preparedness and planning to manage its effects for Australia's defence forces. There's nowhere for politicians to hide.

2. **Public sentiment on the need to act on climate change** transcends demographic, socioeconomic and political elegances and boundaries where action (or inaction) on climate change is unlikely to produce a political differentiation. With the long-lasting impact of the 2019/20 bushfires, it is unlikely we will see a decline in sentiment on climate as we did last decade. This should pave the way for bipartisan political leadership that shifts Australia from a climate change laggard, to a world leader.

The world of climate change

The influence of climate change on our attitudes is influenced by significant environmental catastrophes and developments throughout the world. The decade that was the world of climate change.

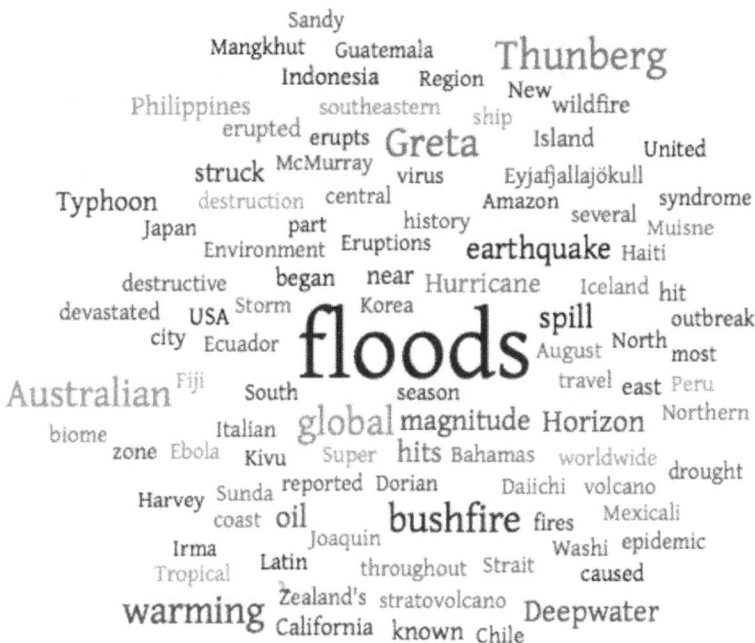

Source: Australia 2030 research Rocky Scopelliti

I'll close this chapter by sharing some of the qualitative thoughts from respondents who ranked climate change as Australia's number one issue and provided commentary on this issue in the coming decade.

IN AUSTRALIAN PROFESSIONALS' OWN WORDS

- *"Australia could be a world leader in renewable energy, but the Government's interest in fossil fuels hampers the progression of this".*

- *"The bushfires are a primary example of distrust in our Government, and their bureaucracy continues to impact how this country operates".*

- *"A sustainable future requires forward thinking policy and action. Repair and preservation of our environment is the biggest long-term issue to face humanity. In the past, Asian countries have not shown leadership, will or practice well enough in this area. No point having growth if our global ecosystem unravels."*

- *"Millennials don't seem to understand the balance needed to action change on a large scale, they would much rather grab onto one thing that needs change and then run to one side of the argument for change or not change. Just look at some of the protests on things like climate change, they run straight to demanded it, even though pretty much everyone agrees we need to do something about this, Millennials don't come to the table with a balance approach wanting to understand everyone's view points and look to find a way forward to get some real change. Climate is one example, I think it's quite scary that Millennials we start to move into more leadership roles, as I have yet to meet one that could be a great leader and inspire change".*

- *"Given our economic makeup, Australia has some of the most to lose from an environmental perspective (depleted resources, wide-spread climate impacts) and actually has a great capacity to invest in the industries given our higher education emphasis. I don't think the level of investment has reflected the importance this could have to Australia".*

- *"The Federal Government does not fund CSIRO and Science enough at all levels and continues to reduce budgets".*

CHAPTER 4

The Asian century

"The world is a funny place, no? Sometimes you pick your dog. Sometimes your dog picks you." (Red Dog)

"As the global centre of gravity shifts to our region, the tyranny of distance is being replaced by the prospects of proximity – Australia is located in the right place at the right time – in the Asian region in the Asian century."

~ Then Prime Minister Julia Gillard's celebrated 2012 launch of the Asian Century White Paper[74]

This word cloud reflects how we felt about our place in Asia Pacific over the past decade.

Source: Australia 2030 research Rocky Scopelliti

In this chapter we will explore the relationship Australia has with Asia. Despite the Gillard government's 2012 Asian Century White Paper[75] proposing the promise of an easy, conflict-free ride on the back of the rapidly industrialising neighbours to Australia's north, together with the promise that economic growth of the region would serve as an insulating factor against potential conflict, we find ourselves at the start of this decade, in an awkward geopolitical sandwich between the United States and China. We will also explore the question of what Australian professionals are expecting in their leaders over the coming decade as they relate to the region. Also, as a Middle Power in the world, how will Australia constructively influence the region as the 'rules-based order' throughout Asia is being superseded by intense competition between the established superpower – United States, and the emerging superpower that is China.

Who are our friends, friends with benefits, and pests in the Asia Pacific region?

Over the past decade, the turbulence of Australia's political leadership, the flattening of global growth and our increasing economic interests centred on Asia have caused many Australians to question our social, economic and cultural place in the world. The past decade brought with it very significant political leadership and policy change throughout the world. Indeed, within Australia, foreign policy that also shaped and influenced the confidence we have in those nations and their leaders.

To understand Australia's foreign policy in context, The Lowy Institute has included in its poll a 'feelings thermometer'[76] that scores on a scale of 0° in temperature (coldest feelings) to 100° (warmest feelings). The thermometer measures Australians' feelings towards 19 countries and territories that have played a prominent role in world events during the year. Exhibit 4.1 includes a subset of those countries where significant favourable and unfavourable movements in temperature were observed over the past decade.

Exhibit 4.1: Feelings towards some countries where 100 = very warm, favourable or 0 = very cold, unfavourable (0°)

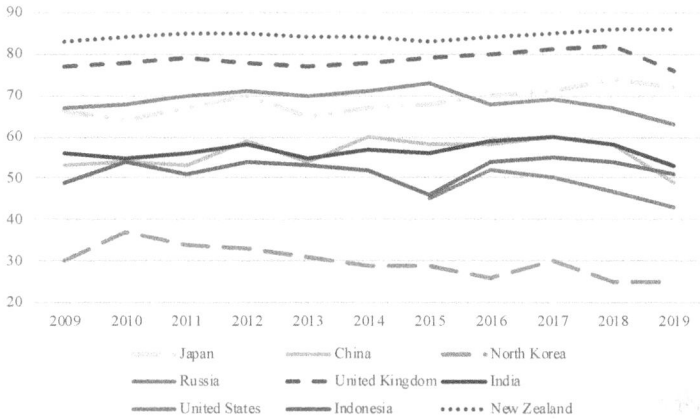

Source: Lowy Institute Poll 2019

Australia's 'besties'. I define these as those countries with a thermometer reading above 61°.

New Zealand – Best friend ever

Top of the chart over the decade was New Zealand. Over the decade, there was a 3° favourable movement that reached its height, a very warm 86° in 2018 and remained so in 2019. I should note that those results reflect the appointment of Jacinda Ardern as Prime Minister of New Zealand in 2017. New Zealand has always received a result above 80° since the Lowy Institute first included it in the poll in 2007. In the 2019 Lowy Poll, New Zealand was also seen by a majority of Australians as our 'best friend'. The impact of the Christchurch massacre is not reflected in the results as the event took place during the fieldwork.

According to Australia's Department of Foreign Affairs and Trade, the economic and trade relationship between Australia and New Zealand is shaped by the Australia New Zealand Closer Economic Relations Trade Agreement (CER or ANZCERTA), which came into effect on 1 January 1983. ANZCERTA is one of the world's most open and successful free trade agreements. Reflecting the high level of economic integration in the TransTasman "single economic market" (SEM) created by CER, two-way merchandise trade in 2018 totalled $17.6 billion (New Zealand was our ninthlargest goods trading partner), while two-way services trade totalled $11.7 billion. In 2018, Australian investment in New Zealand was

$96.7 billion (we are by far their biggest source of foreign investment), while New Zealand invested $47 billion in Australia in the same period[77].

Even though the All Blacks beat the Wallabies 22 times over the decade to the Wallabies 5 wins, our respective women's cricket teams were a little more balanced with both sides winning 10 games each a piece. But on the other hand, Australia's 9 men's cricket test match wins versus 1 to New Zealand over the decade balances the Trans-Tasman rivalry.

New Zealand's Prime Minister Jacinda Ardern tops the list of global leaders, with 88 per cent of Australians confident in her to do the right thing in world affairs (see Exhibit 4.2). Relative to other leaders that is:

- 30 per cent greater confidence than Australian Prime Minister Scott Morrison
- 54 per cent greater confidence than Indonesian President Joko Widodo
- 58 per cent greater confidence than Chinese President Xi Jinping
- 63 per cent greater confidence than US President Donald Trump
- 67 per cent per cent greater confidence than Russian President Vladimir Putin
- 81 per cent greater confidence than North Korean Leader Kim Jong-un.

Exhibit 4.2: Confidence in each leader to do the right thing regarding world affairs 2019 (%)

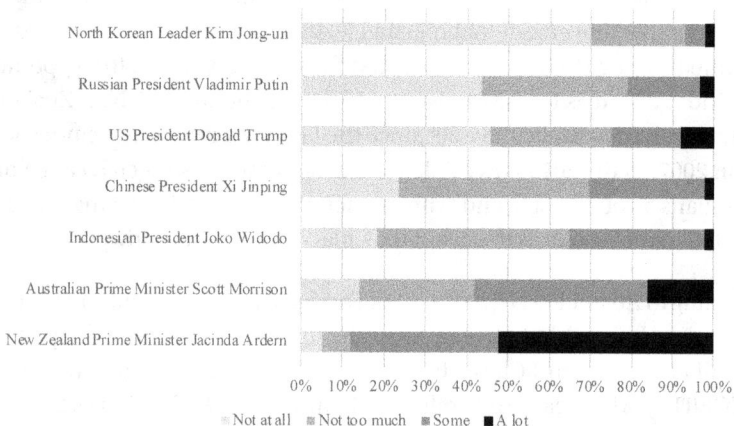

Source: Lowy Institute Poll 2019

United Kingdom – Next best friend.

Like New Zealand, over the decade, there was only a 3° movement that reached its height, a very warm 82° in 2018, however, it took a steep unfavourable fall to 76° a year later. Australia's pessimism about Brexit and the successive change of political leaders may have affected sentiment towards the United Kingdom more generally.

Australia and the UK have a significant and comprehensive relationship underpinned by our shared heritage, common values, strong people-to-people links, closely aligned strategic outlook and interests and substantial trade and investment links. According to Australia's Department of Foreign Affairs and Trade, Australia is in frequent dialogue at the highest levels across government and are likeminded on global issues such as international security, multilateral cooperation and economic issues on the G20 agenda.

The UK is the second largest source of foreign investment in Australia, with the stock of investment valued at $574.8 billion in 2018 (FDI of $98.7 billion). Australian investment in the UK was $408 billion in 2018 (FDI of $118.7 billion). The UK is Australia's eighth largest two-way trading partner, worth $26.9 billion in 2018, and Australia's third largest services trading partner, with Australian services exports to the UK of $5.5 billion and imports of $9.2 billion in 2018[78].

Cricket rivalry between Australia and England has a long history. However, Australia holds the edge in the Ashes series having won on 33 occasions compared to England's 32. The last decade saw the Ashes return to England and Australia each three times. Our women's teams have more draws (3) but Australia tucked away three wins over England's two.

United States – Friend, but in god we trust, not the President

While over the decade, there was a 4° cooling of the feelings towards the US to 63°, in 2019 under Trump's presidency, there was a 10° cooling from 2015 when our feelings were warmest (73°) during the Obama presidency.

In 2018, Australia and the United States marked a centenary of mateship – a friendship first formed in the trenches of the first World War during the Battle of Hamel on 4 July 1918. The two countries maintain a strong relationship, characterised by cultural similarities and robust bilateral arrangements. There are strong formal structures of cooperation between Australia and the United States spanning foreign policy, defence and security, intelligence, development, energy, environment, education, law, trade and investment. The Australia-United States Alliance and the Australia-US Free Trade Agreement (AUSFTA) are central to the bilateral relationship, which also benefits from widespread collaboration across government, academia and business.

According to Australia's Department of Foreign Affairs and Trade, the United States is our largest two-way investment partner, with two-way investment stock reaching $1.6 trillion in 2017. The United States is by far the largest investor in Australia, with investment stock worth $896 billion at the end of 2018. Two-way trade stood at A$70.2 billion in 2017-18[79].

According to the Lowy Institute Poll, trust in the United States 'to act responsibly in the world', has been on a downward trajectory since 2011 whereas half the Australian population (52 per cent) say they trust the United States to act responsibly in the world. The Lowy Institute suggests that these results indicate that Australians' feelings towards the United States, and their support for the US alliance, can be affected by the person who holds the role of US President[80].

Only 25 per cent of Australians have confidence in President Trump 'to do the right thing regarding world affairs', 5 per cent less than for Chinese President Xi.

When it comes to sporting rivalry between Australia and the United States, can it get more nail bitting than swimming? However, let's just say that for our swimmers, the Olympic Games of 2012 in London and Rio De Janeiro in 2016 were a development opportunity with the USA swim team taking away 31 medals in London and 33 medals in Rio, compared to Australia's 10 medals at each Olympic Games.

Japan – Bestie in Asia

Throughout the decade we saw a 5° movement to reach a very warm 72° in 2019, a 9° favourable difference between our feelings for Japan versus the US. According to Australia's Department of Foreign Affairs and Trade, the Australia–Japan partnership is our closest and most mature in Asia and is fundamentally important to both countries' strategic and economic interests. The relationship is underpinned by a shared commitment to democracy, human rights and the rule of law, as well as common approaches to international security.

Australia's Department of Foreign Affairs and Trade acknowledges the Australia-Japan economic relationship is underpinned by complementary strengths and needs. Australia is a safe, secure and reliable supplier of food, energy and mineral resources and a world-class centre for financial and other services. Japan became Australia's largest trading partner in the early 1970s – a position it maintained for 26 years. Japanese investment continues to play a significant role in the development of the Australian economy.

Australia and Japan held the inaugural Ministerial Economic Dialogue in July 2018. The dialogue offers a regular mechanism for high-level engagement on strategic economic and trade cooperation to complement high-level defence and security cooperation and annual leaders' meetings. The dialogue supports the strong and growing trade and investment relationship between Australia and Japan.

Japan was Australia's second-largest trading partner in 2017, with two-way goods and services trade valued at $71.8 billion. Japan is Australia's second-largest export market.

Goods exports to Japan were $45 billion in 2017, which was around 14.9 per cent of our total goods exports. In 2017, Australia's major merchandise exports to Japan included LNG, coal ($16.5 billion), iron ore ($5.6 billion), beef ($2 billion), and copper ores and concentrates ($1.3 billion). Japan was Australia's largest merchandise export market for coal, LNG, beef, aluminium, cheese and curd, liquefied propane and butane, and animal feed. On the other side of the trade ledger, in 2017 Japan was Australia's fourth-largest source of goods imports, including passenger vehicles ($7.6 billion), refined petroleum ($2.8 billion), gold ($2.3 billion), and goods vehicles ($1.6 billion)[81].

The Australia–Japan football rivalry is one that exists between the national association football teams of each country, regarded as one of Asia's biggest football rivalries. The rivalry is a relatively recent one, born from a number of highly competitive matches between the two teams since Australia joined the Asian Football Confederation in 2006. Over the past decade, Japan has shown its strength defeating Australia in the 2011 Asia Cup Final, 2013 East Asian Cup, 2014 International Friendly and 2018 World Cup Qualification with three drawn games. In the national women's competition, Japan leads with five victories against four over the decade.

Australia's 'Warmies'. I define these as those countries with a thermometer reading between 60°–50°.

India

Australians' attitudes towards India have been lukewarm ranging from 56°–53°. But from its height of 60° in 2017, it has sharply declined 7° to 53° in 2019.

India, the world's largest democracy, is a major power. According to Australia's Department of Foreign Affairs and Trade, as strategic partners since 2009, Australia and India enjoy strong political, economic and community ties. India was Australia's eighth-largest trading partner and fifth-largest export market in 2018-19, driven by coal and international education. Two-way goods and services trade with India were $30.3 billion in 2018–19 and the level of two-way investment was $30.7 billion in 2018. We have an ambitious agenda to expand our trade and economic relationship, as outlined in the India Economic Strategy (IES), which was released in July 2018 and endorsed by the Australian government in November 2018[82].

Both countries have shared a very long history and rivalry when it comes to cricket. Over the past decade, India won four of seven series played between the two nations with Australia winning two. While Australia dominated the decade in ICC Women's T20 World Cup wins, this year's final against India attracted a record crowd of 87,174 at Melbourne's MCG to see Australia defeat India. Australia's 4-184 was the highest total in any T20 World Cup final, men's or women's - 23 runs higher than the previous best of 161 compiled by West Indies men at Eden Gardens in 2016[83].

China

Over the decade, the feelings thermometer shows lukewarm attitudes ranging from 53°–49°. However, throughout the decade, the thermometer shows a warming then cooling, then warming and cooling again of attitudes with a significant fall of 11° in the past two years. This corresponds with Australians expressing the lowest level of trust they have ever felt towards China in 2019 at only 32 per cent — a 20-point fall from 2018.

According to Australia's Department of Foreign Affairs and Trade, Australia's bilateral political engagement with China is extensive. Both governments have committed at the highest levels to building on their relationship and strengthening cooperation on important shared interests. At the same time, both sides acknowledge that Australia and China have different histories, societies and political systems, as well as differences of view on some important issues. Australia adheres to its one-China policy, which means it does not recognise Taiwan as a country. Australia maintains unofficial contacts with Taiwan promoting economic, trade and cultural interests.

China is Australia's largest two-way trading partner in goods and services, accounting for 26 per cent of our trade with the world. Two-way trade reached a record $235 billion in 2018–19 (up 20.5 per cent year on year). Our exports to China grew by 23.9 per cent to reach the highest level ($153 billion), driven by demand for Australian iron ore, coal and LNG. China remained our biggest services export market, particularly in education (more than 205,000 students in 2018, an 11 per cent increase year on year) and tourism (more than 1.4 million Chinese visitors in 2018–19)[84].

Chinese investment in Australia is a highly valued and growing part of the bilateral relationship. China is the fifth largest foreign direct investor in Australia ($40.5 billion in 2018), accounting for 4.1 per cent of total foreign direct investment (FDI). In recent years, Chinese investment has broadened from mainly mining to include sectors such as infrastructure, services and agriculture. Australia's foreign investment review framework is established clearly in legislation providing openness and transparency. Australian FDI in China reached $13.5 billion in 2018. Our expertise in banking and wealth management services has seen financial institutions become some of the largest Australian investors in China.

Relations between the two countries have been strained by spying allegations, China's influence in the Pacific, the ban on Huawei technology and other issues. This has recently intensified since the government began canvassing support for a global COVID-19 inquiry.

When it comes to sport, Australian swimmer Mack Horton sparked controversy at the swimming world championships in Korea, refusing to stand on the podium alongside Chinese opponent and accused drug cheat Sun Yang. While Sun sort to broaden the issue from a personal attack to one of disrespecting China, Horton received a standing ovation from fellow competitors in the athlete's village following the protest.

When it comes to leadership, only 30 per cent of Australians have confidence in Chinese President Xi to do the right thing in world affairs. However, confidence in President Xi is still higher than in US President Donald Trump.

Indonesia

Like that with China, over the decade the feelings thermometer shows lukewarm attitudes ranging from 49°–51°. And just like with China, throughout the decade the thermometer shows a warming then cooling, then warming and cooling again of attitudes with a significant fall of 8° in 2015. Indonesia's cool 46° reading in 2015 occurred during the lead-up to the executions of Australians Andrew Chan and Myuran Sukumaran.

According to Australia's Department of Foreign Affairs and Trade, Indonesia – the world's third largest democracy and the nation with the world's largest Muslim population, is one of Australia's most important bilateral relationships. We enjoy an extensive framework of cooperation spanning political, economic, security, development, education and people-to-people ties. Prime Minister Scott Morrison's first overseas visit in his capacity as leader was to Indonesia in August/September 2018 where he and President Widodo elevated relations to a Comprehensive Strategic Partnership (CSP) and announced the conclusion of negotiations on the Indonesia-Australia Comprehensive Economic Partnership Agreement (IA-CEPA).

Two-way investment between Australia and Indonesia was valued at $11.8 billion in 2017, with Australian investment in Indonesia at $10.7 billion and Indonesian investment in Australia at $1.0 billion. Australia's two-way trade with Indonesia was worth $16.8 billion in 2017-18, making Indonesia our 13th largest trade partner. Agricultural products are among Australia's key merchandise exports to Indonesia, while crude petroleum and manufactured goods are key imports. Indonesia was Australia's largest market for wheat ($950 million) and live animals ($575 million) in 2017–18. Two-way trade in services was valued at $5.5 billion in 2017-18. Education-related travel dominates Australian services exports to Indonesia, while our services imports are driven primarily by tourism (much of this is Australian tourists travelling in Indonesia)[85].

When it comes to leadership, only 34 per cent of Australians are confident in Indonesian President Joko Widodo to do the right thing in world affairs.

When it comes to sports, both Indonesia and Australia have made significant advances in football in the past decade with Australia rising to win the 2015 Asian Cup. The Indonesian national team has qualified for the Asian Cup in 2000, 2004 and 2007, however has been unable to move through to the next stage.

Australia's 'Pesties'. I define these as countries with a thermometer reading below 50°.

North Korea

At the opposite and coldest end of the thermometer is North Korea who consistently held that title reaching its lowest point in 2019 at 25°. The destabilising effect within our region and directly aimed at Australia, manifests itself in this sentiment that remained unchanged by the high-profile meetings between North Korea's and the United States' leaders. According to the Lowy Poll, Kim Jong-un receive the fewest votes of confidence from Australians.

According to Australia's Department of Foreign Affairs and Trade, Australia maintains only limited diplomatic relations with the Democratic People's Republic of Korea (DPRK). Australia does not have a trade relationship with North Korea. The relationship continues to be severely constrained by Australia's deep concerns over the DPRK's nuclear, other weapons of mass destruction and ballistic missile programs. Australia has consistently condemned the DPRK's nuclear tests and ballistic missile launches over more than a decade, all of which have destabilised the region and contravened the multiple resolutions on the DPRK adopted by the UN Security Council since 2006.

When it comes to leadership, only 7 per cent of Australians have confidence in North Korean leader Kim Jong-un to do the right thing in world affairs.

Russia

As the second coldest polling at 43°, sentiment towards Russia has no doubt been influenced by the downing of Malaysian Airlines flight MH17, aggression towards the Ukraine and the annexation of Crimea.

According to Australia's Department of Foreign Affairs and Trade, Russia's purported annexation of the Ukrainian territory of Crimea and city of Sevastopol in March 2014, and support of pro-Russian separatists in eastern Ukraine, has overshadowed Russia's relations with Australia and other Western partners. In response, Australia announced targeted financial sanctions and travel bans on individuals and entities instrumental to the Russian threat to the sovereignty and territorial integrity of Ukraine. The United States, European Union, Canada and others have also taken a range of measures, including sanctions against individuals and entities. On 27 March 2014 a majority in the United Nations General Assembly passed a resolution on the Territorial Integrity of Ukraine which emphasised that Russia's purported annexation of Crimea had no validity.

Two-way merchandise trade between Australia and Russia was worth $697 million in 2016, down from $1.837 billion in 2014. Australian merchandise exports to Russia in 2016 were worth $478 million and imports from Russia totalled $219 million. Australian exports to Russia in 2016 included live animals, sugars, molasses and honey, and meat (excluding beef). Crude petroleum dominated Australian imports from Russia in the same period, which also included wood manufactures. Australia's services exports to Russia in 2016 were valued at $111 million and imports of services from Russia were valued at $72 million. Services exports were largely in education-related travel and personal travel[86].

When it comes to leadership, only 21 per cent of Australians have confidence in Russian President Vladimir Putin to do the right thing in world affairs

East? West? or are we caught in the middle?

The escalating trade war, military tensions in the South China Sea, and expanding interest in the Pacific, between the United States and China, place Australia in a very delicate position. Balancing our relationship between two major powers has indeed made Australia the 'meat in the sandwich'. Further, Australia has one of the world's most open free market economies with very few tariffs, however, we are already seeing the escalated targeted use of both non-tariff trade barriers as well as tariffs. As a strong commodity exporter with an ideological aversion to such trade barriers, Australia is regularly being caught in the middle.

According to the Lowy Institute, 50 per cent of Australians agree that the Australian government should 'put a higher priority on maintaining strong relations with the United States, even if this might harm our relations with China'. However, the other half, 44 per cent, think Australia should 'put a higher priority on building stronger relations with China, even if this might harm our relations with the United States'. It has also divided the country demographically on foreign policy in terms of where the government should put its priorities. Younger generations (18–29 years) favouring 'building stronger relations with China (55 per cent), even if that meant harming our relations with the United States' versus 62 per cent of 60+ years, favouring 'maintaining strong relations with the United States, even if this might harm our relations with China'[87].

Despite that, when it comes to the question of which of the two nations Australians trust to 'act responsibly in the world', its crystal clear. In 2019 only

30 per cent of Australians trust China, a massive 20 per cent decline since 2018 compared to 52 per cent for the United States. Australians' trust in China has fallen to its lowest level in 15 years. This result follows a turbulent period in Australia-China relations associated with political donation scandals and banning the Chinese technology company Huawei from providing broadband and 5G technology. According to news reports, Prime Minister Scott Morrison said he isn't 'waiting by the phone' for an invitation to visit China, and he said people need to be 'very wide-eyed' about China's behaviour, including its crackdown on the protests in Hong Kong, but he was "constantly surprised at the surprise" surrounding Beijing's rise[88].

Our trust in the US and China— 'It's business, not personal'.

China and the US are our number one and two trading partners. Our economic prosperity is deeply intertwined in a delicate balance between two superpowers that are economically at war with one another. While Australians know that the US presence in Asia over the past three-quarters of a century has underpinned regional stability and prosperity, we aren't so comfortable with the idea of the region being dominated by China. We prefer a balance of forces in Asia that includes the US, with a general acceptance of international norms and the rule of law. But trust in the US has fallen sharply by 31 per cent over the decade to a record low of 52 per cent— when Barack Obama became President. Our trust in China to act responsibly has been in gradual decline since 2011 also, but in 2019, it took a sharp decline of 20 per cent to 32 per cent (see Exhibit 4.3).

Exhibit 4.3: How much do you trust the following countries to act responsibly in the world? (%)

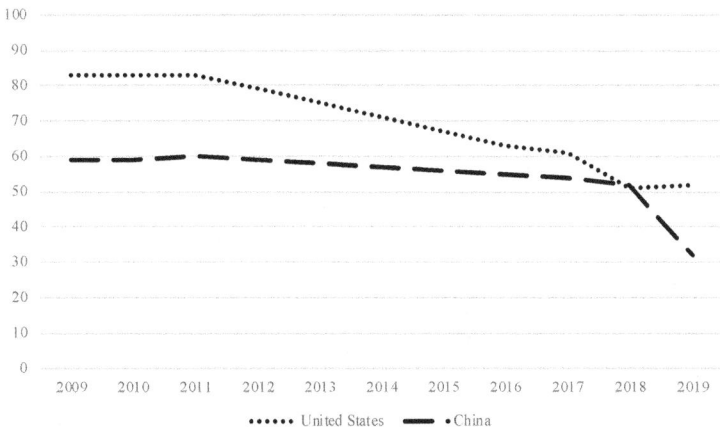

Source: Lowy Institute Poll 2019

At a leadership level, only 30 per cent of Australians have confidence in President Xi to do the right thing in world affairs compared with President Trump at 25 per cent (see Exhibit 4.2). However, despite the lack of trust and confidence in leadership of both superpowers, according to the Lowy Institute poll, Australians overwhelmingly believe that it's possible for Australia to have a good relationship with China and a good relationship with the United States at the same time (81 per cent).

As can be seen in Exhibit 4.4, 45 per cent of Australian professionals expect 'Values led policy change including integrity, honesty, transparency, humility and accountability' when it comes to the leadership qualities, they believe will be the most important to world leaders over the next 10 years. Twenty-two per cent nominate the 'pursuit of both the creation of social and business value' as being the most important.

Exhibit 4.4: What leadership qualities do you believe will be the most important to world leaders over the next 10 years? (%)

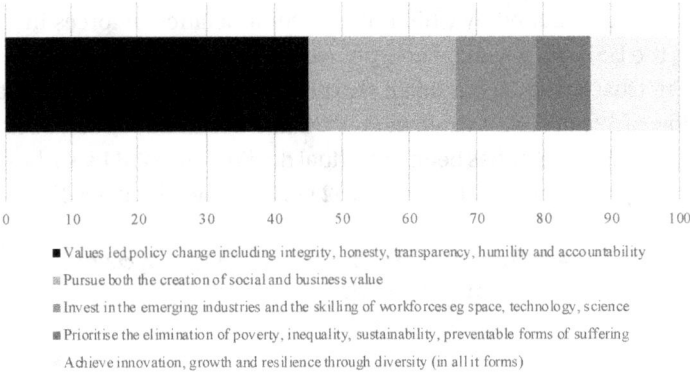

■ Values led policy change including integrity, honesty, transparency, humility and accountability
■ Pursue both the creation of social and business value
■ Invest in the emerging industries and the skilling of workforces eg space, technology, science
■ Prioritise the elimination of poverty, inequality, sustainability, preventable forms of suffering
Achieve innovation, growth and resilience through diversity (in all it forms)

Source: Australia 2030 research Rocky Scopelliti

When we look at those respondents who ranked 'values led policy change including integrity, honesty, transparency, humility and accountability' as number one and then look at their 'opinions, on which issues they believe will affect the world and Australia over the next 10 years', we uncover some fascinating insights. For example, 'climate change and the environment', 'low economic growth' emerge as the most significant issues for both Australia and the world by more than half of those respondents (see Exhibit 4.5). This reflects the sustainability concerns by Australians as it relates to the dependency on Asia for growth.

Exhibit: 4.5 In your opinion, which of the following issues will affect Australia (the world) over the next 10 years? (%)

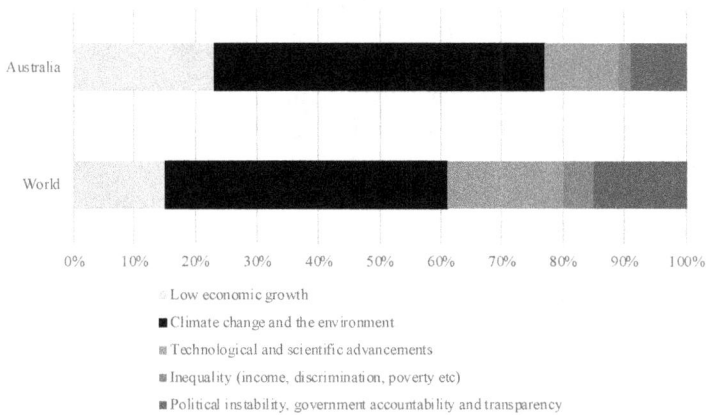

Source: Australia 2030 research Rocky Scopelliti

In 2020, Asian economies are predicted to become larger than the rest of the world combined – are we ready for the Asian century?

The re-emergence of Asia is among the most important shifts in the world's centre of gravity that will occur in our lifetimes. It's been said that in the 19th century, the world was Europeanised, in the 20th century, it was Americanised and now, it is being Asianised – and much faster than we may think. For example, three of the world's four economies are in Asia, and the fourth is the United States.

One of the many fascinating aspects about the rise of Asia is its intraregional dynamics. Half of the world's population and middle class call the region home and Asia accounts for around one-third of global trade in goods. By 2040, it is predicted to generate more than 50 per cent of world GDP and account for as much as 40 per cent of global consumption[89]. Intraregional networks are also driving progress. Around 60 per cent of Asian countries' total trade in goods occurs within the region, facilitated by increasingly integrated Asian supply chains. Intraregional funding and investment flows are increasing, with more than 70 per cent of Asian start-up funding coming from within the region. Flows of people – 74 per cent of travel within Asia is undertaken by Asians – help to integrate the region as well.

As outlined in the Australia 2030 research in chapter 3, 'Climate change,' 'Low economic growth, 'Technological and scientific developments' are the major issues Australian professionals perceive will impact the world. While Asia has benefited enormously from globalisation, it also encapsulates many of the world's problems such as the emergence of COVID–19 in China, and the manner upon which many nations in the region managed that crisis.

In 2020, the Financial Times reported that in purchasing power parity (PPP) terms, Asian economies have been forecasted to become larger than the rest of the world combined for the first time since the 19th century[90]. This inflection point will not just see Asia growing richer; as it becomes more integrated, it has become the engine room for global growth (see Exhibit 4.6.

Exhibit 4.6: Share of world GDP at PPP (%)

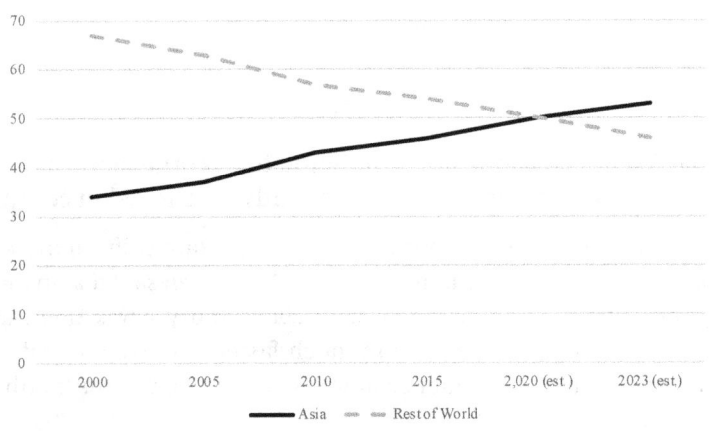

Data Sources: FT, IMF

While China and India explain large parts of that trend, the economic growth fuelled by other economies also provide explanations for that trend. For example, Indonesia is on track to become the seventh-largest economy by 2020, predicted to overtake Russia by 2023. China is now a bigger economy in PPP terms relative to the United States and India has emerged as the third largest economy with a GDP about double the size of Germany or Japan.

Asia's remarkable rise and intraregional integration has seen the mutual benefits of cooperation become ever more apparent. For example, the revival of China's relationships with India, Japan, and South Korea, as well as the recommencement of the China-Japan-ROK trilateral summit. Regional cooperation platforms such as APEC, ASEAN and the Shanghai Cooperation Organization are gaining much more momentum.

Asia is now at the heart of the action for multilateral trade liberalisation. The reformed Trans-Pacific Partnership (TPP) (also known as the Comprehensive and Progressive Agreement for Trans-Pacific Partnership CPTPP) has been revived under Asian leadership and came into force at the start of this year. This was after the newly elected US president Donald Trump withdrew the US signature from TPP in January 2017. Talks are also progressing towards the Regional Comprehensive Economic Partnership (RCEP). This is a proposed free trade agreement in the Asia-Pacific region between the 10 member states of the Association of Southeast Asian Nations (ASEAN) being Brunei, Cambodia, Indonesia, Laos, Malaysia, Myanmar, the Philippines, Singapore, Thailand, Vietnam and free trade agreement partners being Australia, China, New Zealand and South Korea. India, which is also an ASEAN free trade agreement partner, opted out of the RCEP in 2019.

These agreements will continue to evolve and likely attract new members, providing a flexible, multi-track path to economic integration in Asia. For example, the more rigorous CPTPP can help to set standards for future trade for advanced economies, while the less-demanding RCEP will offer a way for developing countries to participate in free trade. When these come together with infrastructure initiatives, trade pacts and other groupings, they will boost pan-Asian coordination on connectivity and trade liberalisation. In turn, this can help fuel a virtuous cycle of mutual gain and closer integration.

A more integrated Asian community – one that brings together developed and developing countries and various economic systems – could provide the momentum and modalities to reinvigorate multilateralism at the global level. Solutions that have been adapted to Asia's diverse conditions may well prove to be useful templates for the rest of the world.

Geopolitically, we find ourselves, socially, economically, culturally and politically divided and unclear on the impact of the Asian century and our relationship to it. This division is highlighted in the Australia 2030 research through respondents' beliefs on whether 'the world will be better economically, socially, culturally and politically better off led by Asia, then the past decade that was led by Europe and the United States of America (see Exhibit 4.7).

The results indicate that Australian professionals are unsure whether the world would be better off which reflects the sentiment and concerns by the Lowy Institute study in relation to leadership and trust.

Exhibit 4.7: Q. 'Analysts predict that in 2020, Asia is set to become the largest economic region. In your opinion, over the next decade, will the world be better economically, socially, culturally and politically lead by Asia, then the past decade that was led by Europe and the United States of America? (%)

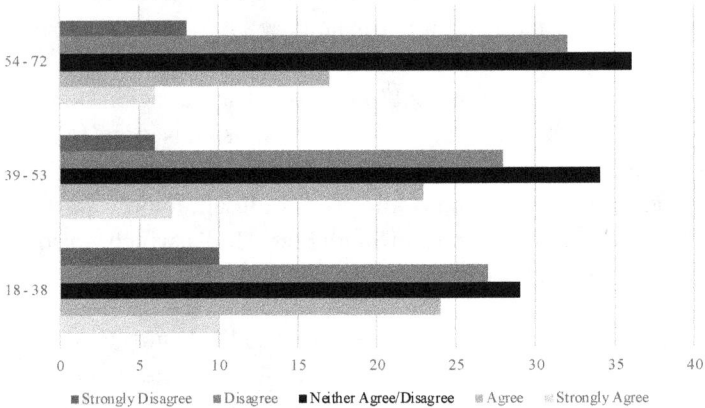

Source: Australia 2030 research Rocky Scopelliti

Where and what are we in the region?

Despite those reservations, Australia plays an integral role in the region as power and wealth shift eastward into the region. To measure that influence, the Lowy Institute created a power index, which comprises indicators associated with economic resources, military capability, resilience, future trends, diplomatic influence, economic relations, defence networks and cultural influence[91]. The index ranks 25 countries (see Exhibit 4.8). For 2018 the index indicates that:

1. The United States remains the preeminent power in Asia
2. China, the emerging superpower, is rapidly closing in on the United States
3. Japan and India share major power status
4. North Korea, Russia are misfit middle powers in Asia
5. Singapore, Australia and South Korea are over performers in the region.

Exhibit 4.8: Asia Power Index 2019

Rank	Country/Territory	2019 Score	Overall Trend	Economic Resources Trend	Military Capability Trend	Resilience Trend	Future Resource Trend	Diplomatic Influence Trend	Classification
1	United States	84.5	-	↓	-	-	-	↑	Superpowers
2	China	75.9	↑	↑	↓	-	↑	↑	
3	Japan	42.5	-	-	↓	↓	-	↑	Major powers
4	India	41.0	-	-	-	-	↓	-	
5	Russia	35.4	↑	-	↑	↑	↓	↑	Middle powers
6	South Korea	32.7	-	-	-	-	-	-	
7	Australia	31.3	-	-	↑	↓	-	↓	
8	Singapore	27.9	-	-	-	↓	-	↓	
9	Malaysia	22.8	-	-	↑	-	-	↑	
10	Thailand	20.7	↑	↑	↑	-	-	↑	
11	Indonesia	20.6	↑	-	↑	↓	↓	↑	
12	New Zealand	19.9	-	-	↑	↓	-	↑	
13	Vietnam	18.0	↑	-	↑	↓	-	-	
14	Taiwan	15.9	↓	↑	↓	↓	-	↓	
15	Pakistan	15.3	-	-	↑	↓	↓	-	
16	North Korea	14.0	↑	-	↑	↓	↑	↑	
17	Philippines	13.7	↑	↑	↑	-	-	↑	
18	Bangladesh	9.7	↑	-	↑	↓	-	↑	Minor powers
19	Brunei	9.1	-	-	-	-	-	↓	
20	Myanmar	8.9	↑	-	↑	-	-	↓	
21	Sri Lanka	8.5	-	-	↑	↓	↓	↑	
22	Cambodia	7.7	↑	-	↑	↑	-	-	
23	Laos	6.4	↑	-	↑	↑	-	↓	
24	Mongolia	6.2	-	-	↑	↓	↑	↑	
25	Nepal	4.7	↑	-	↑	↓	-	↑	

Source: The Lowy Institute

"In a stable ecosystem, smaller species thrive. In a dog-eat-dog world, everything starts to look like dinner."[92]

~ Edward Luce, journalist, Financial Times

According to researchers at the Lowy Institute, it will be the 'Middle Powers' that will determine the future of Asia. This thinking is based on the expectation of a prolonged period of competition for power as the US 'rules based' order declines and tensions surround the rise of 'power-based rules'. The idea is based on three principles:

1. **The rules-based order** in Asia is being superseded by intense competition between the United States – an established superpower – and China – an emerging superpower. An example of this is China asserting itself in the South China Sea ignoring international conventions of the Law of the Seas and determinations by the UN Permanent Court of Arbitration. Since that determination in 2016, China has continued to 'redraw' geographical boundaries through its unlawful land reclamation around contested features of the South China Sea – developing military facilities on some of these features, including the installation of powerful surface-to-air missiles.

The United States defence of this 'rules-based order' has been contradicted through its withdrawal from the Trans-Pacific Partnership, the Paris Agreement on climate change and the Human Rights Council. The unilateralist approach to foreign policy in confronting China through accords such as the 15 January 2020 US-China Phase One Trade Agreement only serves to work against the World Trade Organisation.

2. **Minor and Middle Power countries** have the most to lose from a transition from rules-based to power-based order. For example, Asia is now at the heart of the action for multilateral trade liberalisation and has benefited from interregional integration of their trade supply chains. Rules and regional stability have allowed the Minor and Middle Power countries to compete evenly. However, when rules are not consistently upheld and adhered to, Minor Power countries become vulnerable to Major Powers who seek to exploit them for their geopolitical gain. Examples of this include funding programs involving loans by China to Pacific Islands.

3. **Collectively, Middle Powers provide** a stabilising balance of power in the region. As can be seen in Exhibit 4.8, Asia has many Major, Middle and Minor Power nations who are interregionally integrated and rely upon one another for their economic, social, cultural prosperity. Despite their disinterest in taking sides between the US and China, they assert themselves on an issue by issue basis. Middle Powers do, collectively, wield economic clout. In current GDP terms the Asian Middle Powers collectively control a GDP of USD8.7 trillion — twice that of Japan, three times that of India and two thirds that of China.

The conundrum for Australia is between the question of who rules in Asia and who writes the rules for it? Gideon Rachman, the chief foreign affairs commentator of London's Financial Times, provides insights into that question in his book *Easternization: Asia's Rise and America's Decline*. The book is centred on the themes of Asia up, America down, Europe out and the shift of power to the Pacific. Rachman is concerned with the notion that America is losing its grip on world affairs which is "in danger of becoming conventional wisdom". Rachman thinks that if the United States has the will then it has the resources to stay near the top of the global rules game.

Politically, Australia has been more restrained than the US over its concerns about China. However, Australia is just as vexed about what sort of ruler China aims to be. For example, Australia's 2016 Defence White Paper raised repeated concerns about the need for international rules. i.e. using the word "rules" 64 times, 48 of them in the formulation "rules-based global order." What does

Australia face in Rachman's era of Easternisation. Rachman argues Australia "faces an acute strategic dilemma," even as it greets "the rise of Asia with exuberant enthusiasm, treating it as an unparalleled opportunity to secure Australia's prosperity long into the future." The dilemma he presents is that Australia and New Zealand "risk becoming isolated Western outposts, cut off from its political and cultural hinterland. As a result, the vision of China asserting its influence across the South China Sea and in Southeast Asia set off alarm bells in the Australian elite."

The fear of a coercive China bending Southeast Asia and the Pacific Islands to its will has driven Australia to change its definition of the region from the Asia-Pacific to the Indo-Pacific. "The notion of the Indo-Pacific emphasises India's importance and challenges the idea of a region that inevitably revolves around China," says Rachman. The potential for the risk of China to exploit the current COVID-19 crisis to the Pacific Islands through assistance packages was reportedly elevated by the Prime Minister Scott Morrison's appeal to the G20 leaders during an emergency video meeting. Some 80 nations were reported to have already appealed to the International Monetary Fund for assistance. The G20 held concerns about the prospect of nations collapsing under the economic and health strains caused by the COVID-19 crisis[93].

Australia confronts a rapidly changing Asian region. According to Australia's Foreign Policy White Paper[94] "In this dynamic environment, competition is intensifying, over both power and the principles and values on which the regional order should be based." That sentiment was certainly reflected in the divide between respondents on whether the world will be economically, socially, culturally and politically led by Asia over the coming decade, overall sentiment is positive (56 per cent).

KEY POINTS

- Over the past decade, the turbulence of Australia's political leadership, the flattening of global growth and our increasing economic interests centred on Asia has caused many Australians to question our social, economic and cultural place in the world.
- Besides our relationship with New Zealand, the UK and the United States, over the past decade Australians have felt an ever-increasing warmth in our relationship with Japan – our best friend in Asia.
- While Australia's warmth relationship with China, India and Indonesia has remained relatively stable over the past decade, that warmth has declined over the past two years. The challenge for Australia over the coming

decade will be to see these relationships shift in temperature from *warmth to besties.*

- China and the US are our number one and two trading partners. Our economic prosperity is deeply intertwined in a delicate balance between two superpowers that are economically at war with one another. While Australians know that the US presence in Asia over the past three-quarters of a century has underpinned regional stability and prosperity, we aren't so comfortable with the idea of the region being dominated by China.
- The re-emergence of Asia is among the most important shifts in the world's centre of gravity that will occur in our lifetime. It's been said that in the 19th century, the world was Europeanised, in the 20th century, it was Americanised and now, it is being Asianised – and much faster than we may think. Today, the top three world economies are in Asia. By 2040, Asia is predicted to generate more than 50 per cent of world GDP and account for as much as 40 per cent of global consumption.
- Geopolitically, Australia finds itself socially, economically, culturally and politically divided and unsure of the impact of the Asian century. This division is highlighted in the Australia 2030 research through respondents' beliefs on whether 'the world will be better economically, socially, culturally and politically better off led by Asia, then the past decade that was led by Europe and the United States of America.
- Despite those reservations, Australia plays an integral role in the region as a Middle Power as the transition eastward occurs. It will be the Middle Powers that will determine the future of Asia. This thinking is based on the expectation of a prolonged period of competition for power as the US rules-based order declines and tensions surround the rise of power-based rules from China. Middle Powers provide a stabilising balance of power in the region. Asia has many Major, Middle and Minor Power nations who are interregionally integrated and rely upon one another for their economic, social, cultural prosperity.

We have now crossed two major tipping points when it comes to the question of Australia's relationship and role with Asia that will set the agenda for the coming decade.

TIPPING POINTS

1. **The Asian century has arrived.** Asia is now at the heart of the action for multilateral trade liberalisation, which the rest of the world will significantly benefit from.
2. **As a Middle Power, Australia, together with other Asian Major, Middle and Minor Powers, can provide regional stability.** Asia has many Major, Middle and Minor Power nations that are interregionally integrated and rely upon one another for their economic, social and cultural prosperity. Despite their disinterest in taking sides between the US and China, they assert themselves on an issue by issue basis stabilising influence.

Asia Pacific's richness and diversity

Throughout the decade there were significant social, cultural, economic, technological, scientific and environmental challenges throughout Asia Pacific. This word cloud thematically reflects the decade that was Asia Pacific.

Source: Australia 2030 research Rocky Scopelliti

I'll close this chapter by sharing some of the qualitative quotes from the 337 respondents who provided explanations for their choices.

IN AUSTRALIAN PROFESSIONALS' OWN WORDS

- *"Chinese influence in 'buying' up Pacific nations via gifting infrastructure is a Trojan Horse into Australia."*

- *"The population of the globe will not tolerate China and the remaining Asia's to rule by dictatorship. They have barely tolerated North Korea as it is."*

- *"It is clear that Asia will take a far more dominant place in influencing the world and it is yet to be seen how the China influence will play out. Will they support a rules-based order for the world to play by recognising it is likely to be different to the decades post WW2 or will they contribute in a far more individualistic, aggressive style. This will define whether the world is better or not."*

- *"I agree that Asia and the APAC region will be driving economic and technological advances over the next decade due to a vast of highly skilled labour force entering university and the lower cost to export/import products and services to drive the productisation economy towards extremely different industries. Due to its geographical location it also makes it an ideal innovation and transportation hub for the APAC region to diversify their products offerings due to low transportation cost and business models."*

- *"Economically we may be better but approaches to politics (non-democratic) and resulting social and cultural impacts could be worse."*

- *"Lots of unknowns. Big risk is the rise of authoritarianism - geo-political issues related Chinese positions such as OBOR, Chinese disregard for international laws (e.g. South China sea issue) - But opportunities are China and India with their size of population again gaining economic status they held once."*

- *"India - will be the gold country. Just wait and watch. They are heading the right direction."*

- *"Globalization enabled Asian countries receive funds through investments and job opportunities, which improved the poverty, standard of living and overall a matured, skilled and educated workforce. This workforce is now trying break the barriers and innovating at a rapid pace."*

- *"Rising nationalism across the globe is affecting global growth. China (in the Asia region) is becoming more so and willing to spread its 'rightful place in the sun.' Efforts to subjugate Taiwan / Hong Kong and dominate smaller regional nations (road and belt initiatives) financially will further destabilise global markets. Especially given existing US policy (god forbid if Trump gets re-elected) and EU becoming more insular."*

- *"Asia has a greater likelihood of regional state-based conflict (involving China be it, via South China Sea, Hong Kong/Taiwan, North Korea, Philippines) and will have greater affliction of issue motivated unrest (Indonesia, Philippines, Western China). I generally disagree that the 4 of the 5 global communist countries located in Asia are better to lead socially/culturally/political along with the fact that there are too many competing interests regionally for it to be a stable region to lead the world. The US will remain an economic, cultural and military leader past this decade with Europe will regulating itself into fracturing further past Brexit."*

- *"I think that the rise of China particularly (but perhaps also India) will provide a balance against US dominance and hopefully lead to the US taking unilateral military action as has been the case in the Middle East over the last decade. The counterbalance will hopefully be that US pressure on China will encourage China to gradually liberalise their approach to human rights etc. Being an optimist, I am ignoring the possibility of a major war between China and the US."*

- *"A challenging question to answer. Perhaps better economically, although significant political and cultural differences may well have detrimental impact that will challenge successful change and growth."*

- *"As the economic power of communities rise, the opinions and ethics of their individuals evolve. Through better communications we have the ability to shape the ethics and values of the rising economic and social leadership cadre across Asia. Of course, it is up to us to decide whether to do that positively or negatively."*

CHAPTER 5

Demographics

"Tim Tam?"
(Muriel's Wedding)

"Hello possums!"

~ Dame Edna Everage, (Barry Humphries)

This word cloud represents who we are demographically and the major social changes of the past decade.

Source: Australia 2030 research Rocky Scopelliti

The 2010s was the decade in which we saw significant demographic change in Australia and indeed the world. We know that populations are ageing, and that life expectancy has steadily been increasing by two to three years each decade. At the other end of the age spectrum, Millennials have now become the largest demographic group on the planet representing one in three (2.1 billion) people[95]. They are likely to be the first generation to have a 50 per cent chance of living to 100 years[96]. This means they are also likely to see the emergence of four-generation families and so our notion of family structure will profoundly change from what we've ever known. With their proportionate representation in society – whether as leaders in business, government or institutions, be they spiritual, academic, scientific or technological – their influence will only increase from here on. Millennials are integral to considering the decade ahead.

Demographers Neil Howe and William Strauss invented the label Millennials, but the demographic cohort is also commonly referred to as Generation Y, those born between 1981–2000. Social researchers creatively applied that label as it follows Generation X, those born between 1965–1980. 'Baby Boomers' is the label used to describe those born between 1946–1964, most of whom are the parents of Millennials.

At a global level, there are three very significant demographic trends that will create significant impacts in Australia.

1. **There will be about 1 billion more people, and they will live longer.** The world should reach 8.5 billion people by 2030, up from 7.3 billion in 2015. The fastest growing demographic will be the elderly, with the population of people over 65 years old at 1 billion by 2030. Most of those new billion will be in the middle class economically, as the percentage of citizens in dire poverty continues to drop. Even as the middle swells, however, the percentage of all new wealth accruing to the very top of the pyramid will continue to be a major and destabilising, issue.
2. **Two-thirds of people will live in cities.** The urbanisation of our populations will increase, creating more megacities as well as small- and medium-size metropolises. The effects will include the need for more big buildings with better management technologies and we will need more food moved in from where we grow it to where we eat it — or rapidly expand urban agriculture.
3. **Generational polarisation:** For the first time ever, there are as many people over the age of 30 as under the age of 30. This tipping point has

profound implications for the global economy. From now on, the world will have ever fewer poor people, about the same number of youth and young adults, and many more old and wealthy people. Over the next decade, the world will add 800 million people above the age of 30, and there will be 100 million less people under the age of 30.

Let's begin by looking at a snapshot of the Australian population.

Population

Australia's population is growing annually more than ever before. Since 2010 the population has grown 15 per cent and is predicted to grow by 10.5 per cent to 2030[97]. The past decade has seen the urban concentration of Australians increase to 85.9 per cent of the population. While the past decade saw this concentration increase by 0.9 per cent, the coming decade to 2030 will see this increase double to 1.9 per cent, and to 87.8 per cent. This growth will likely result in a continued increase in housing density, rising house prices and increased investment in infrastructure and services particularly in major cities. Today Australia ranks number 55 in the list of countries by population and that ranking will decrease to 59 by 2030[98].

Cultural diversity

Australia has evolved into a nation of people from more than190 countries and 300 ancestries and is now one of the most culturally diverse nations in the world. Migration continues to play an important part of Australia's population growth with migrants representing approximately 47 per cent. Looking ahead to 2030, the proportion of migration to the population growth is predicted to increase by approximately 2 per cent[99].

Fertility rates

Australia's fertility rate declined over the decade by 0.2 births per women. Australia is experiencing record low fertility rates (1.74 children per woman) in 2019 but record high number of births overall (315,147 a year) At the same time, mums are becoming older. In 1998 the median age was 29.5. Two decades later, the age of mothers at the time of giving birth increased to 31.4 years[100].

Marriage rates

Despite the fact that the Australian population had doubled since the early 1970s, in 2017 there were fewer marriages than in 1970. There were 116,066 marriages in 1970, and only 112,954 in 2017. In 1971 the median age for men and women to marry was 23.4 years and 21.1 years respectively. In 2017 the median age of men and women marrying was 30.4 years and 28.8 years respectively. Most couples now live together before getting married. In 1975, only 16 per cent of marriages were preceded by cohabitation. By 2017, 81 per cent were. The year 2000 also saw an inflection point where there were more marriages performed by civil celebrants than by ministers of religion. By 2017, 78 per cent of marriages were performed by civil celebrants[101].

Workforce

Approximately 13 million Australians are employed, of which 8.9 million are employed full time and 4.1 million employed part time. Gen Y and Z now comprise the majority of the workforce[102]. While these generations (born since 1980) make up more of the workforce than those born before 1980, these new generations are also moving into leadership roles. The average job tenure is 2.9 years. That means the emerging generations are expected to have 18 jobs over 6 careers in their lifetime. We are now in an era of employment flexibility and empowered workers.

Australian Gen Zs (born around 2000)

In 2003, the first nationally represented longitudinal study of Australian child development was established and published by the Australian Institute of Family Studies[103]. This study provides a window into understanding children's development within Australia's social, economic and cultural society. In 2019, the study found that:

- *Physical health:* Only one in seven Australian children met the Australian guidelines for physical activity of at least 60 minutes per day. Nine out of ten children (92 per cent) were not meeting the recommendation of eating at least five serves of vegetables per day
- *Sleep:* A quarter of 12–13-year olds and a half of 16–17-year olds were not meeting the minimum sleep guidelines (9–11 hours) on school nights. Those that weren't getting enough sleep were more likely to show symptoms of poor mental health (e.g. anxiety, depression and unhappiness).
- *Romance:* By the age of 16–17, around two thirds of teenagers had been involved in a romantic relationship and around a third had had sex. Almost half of girls and one third of boys aged 16–17 years reported experiencing some form of unwanted sexual behaviour towards them within the past year.

- *Risky driving:* Almost 55 per cent of learner drivers and 80 per cent of P-platers aged 16–17 had engaged in some form of risky driving on at least one of their 10 most recent trips. About 10 per cent of teens had been the passenger of a driver who was 'under the influence' during the past year.
- *Education interests:* Boys were more likely to choose Advanced Maths, Engineering, Physics, Technology, Business and Finance subjects than girls. Girls were more likely than boys to select Biology, Creative Arts, Health, Psychology, Legal Studies and Society and Culture.
- *Volunteering:* Approximately 40 per cent of 12–13-year olds and 50 per cent of 16–17-year olds reported volunteering in the past 12 months.
- *Family concerns:* More than half of 10–11-year olds were worried about fighting in their family. Serious family illnesses, terrorism and the environment were also major issues they worried about.

What is fascinating is that this study reflects the impact of the significant and profound changes that have arisen over the past two decades. For example, technologically, this generation would not know what life was like to not be connected to their devices, all the time, anywhere, anyplace. To them, being connected is as natural as the air that they breathe. But that always connected access to the world around them, doesn't necessarily translate into better health outcomes such as dietary or mental health. Sadly, youth suicide is the leading cause of death among young Australians[104].

For the past 18 years, Mission Australia has been running a Youth Survey. The 2019 Youth Survey[105] involving 25,126 participants aged between 15–19 years revealed many concerns. For the third year running, the study found that mental health is the top national concern for young people, with other personal concerns relating to mental health, such as coping with stress, school or study problems and body image. Young people also reported that they feel they don't have enough of a say on important public affairs issues. The growing public dialogue and experience of issues such as extreme weather events and drought are clearly affecting young people's view of the world. These results taken together indicate that young people in Australia feel disenfranchised and deeply concerned. The report suggests that this perceived inability to have their voices heard through formal channels is perhaps causing them to engage in informal ways to get heard, such as climate strikes.

Australian Millennials, their Baby Boomer parents and their collective Youthquake

My last book, *Youthquake 4.0 – A Whole Generation and the New Industrial Revolution*, detailed the impact of the Millennial generation under the umbrella of the word 'Youthquake'. That word became the 2017 Oxford Dictionary Word of the Year[106] and is defined as a 'significant cultural, political, or social change arising from the actions or influence of young people'. Surprisingly, this is not a new term. Youthquake was first coined by *Vogue* magazine's editor-in-chief Diana Vreeland in 1965 to describe the cultural movement on the streets of London by a new generation of young people we now know as Baby Boomers[107]. Vreeland wrote in her article entitled Youthquake – "Youth is surprising countries east and west with a sense of assurance serene beyond all years. First hit by the surprise wave, England and France already accept the new jump off age as one of the exhilarating realities of life today. The same exuberant tremor is now coursing through America, which practically invented this century's youth in the first place".

Ironically the term 'renaissance' five decades later has been used to describe Baby Boomers' children – the Millennials. We shouldn't be surprised that the first and most powerful influence on Millennials was their parents. Youthquake for Baby Boomers was so well captured in the lyrics "There's a whole generation, With a new explanation, People in motion" in Scott McKenzie's 1967 hit single song and generational anthem 'San Francisco (Be Sure to Wear Flowers in Your Hair)'[108]. That song reached number one in the United Kingdom encapsulating the spirit of a generation during the 1960s who were craving significant cultural, political and social change.

Are Millennials an unusual generation? A problematic generation? A puzzling generation? A preoccupied generation? An entitled, ungrateful generation? Was there ever a rising generation in history not given those labels? As George Orwell so well-articulated "each generation imagines itself to be more intelligent than the one that went before it, and wiser than the one that comes after it".

Or are Millennials simply a generation that, like their Baby Boomer parents, have shaped their social, cultural and economic beliefs based on the environment in which they grew up? For example, might Millennials' perceived disloyalty reflect their perfectly understandable need to explore life's options? Might their perceived entitlement, be a misinterpretation of their empowerment? Might their pre-occupation with technology reflect their desire to remain connected with their own relationships within their communities? The tsunami of stereotyping and typecasting directed at Millennials would seem to be overgeneralisations born out of misinterpretations, since they seem to be clearer than many older generations about how to live in the contemporary world and they have a vision for its future. To them, the world is their neighbourhood.

It's worth reflecting on one of the many essential points about their Baby Boomer parents to help explain their influence on Millennials. Baby Boomers were shaped by two different and contradictory influences on their view of the future.

First, Baby Boomers grew up in the post-war economic boom of the 2nd Industrial Revolution where electric power was used to achieve mass production and the division of labour. There wasn't just a baby boom, but also manufacturing, mining and housing booms. That period also saw the rise of the working class. Baby Boomers were enveloped in prosperity developing an unquenchable thirst for in-home appliances, telephones fixed to walls, white goods, televisions, motor vehicles and leisure activities that fuelled the creation of many new consumer markets.

The era ushered in suburbia. New neighbourhoods sprang up and the locality symbolised status, class and lifestyles. To Baby Boomers, life was a never-ending pathway of gratification, with the promise of success, wealth and opulence.

This was spoilt by the threat of no future at all.

The second powerful but contradictory influence on them was the Cold War. They were growing up in the era of mutually assured destruction. This was the era of nuclear weapons being massively stockpiled by the USA and the then USSR, and the threat they would be used, triggering world war three. Baby Boomers grew up with tension points in countries around the world that were backed by the two superpowers.

So how did they reconcile these contradictory influences? Their motivation shifted from delayed gratification, to embracing *instant* gratification despite this description being synonymous with Millennials. Baby Boomers' generational catchcry was 'We're not here for a long time; we're here for a good time'. They became famous for their impatience—rushing into marriage, rushing into parenthood, rushing into debt, for their frivolous spending, and for their reluctance to plan for the long term.

Baby Boomers were labelled the 'Me Generation' by their parents—the

Silent Generation born between 1927–1945—who were puzzled by their self-indulgent, live-for-now mentality and liberated sexuality brought about by the contraceptive pill. Baby Boomers were living in a way that was consistent with those contradictory influences.

As history now records, Baby Boomers were thankfully here for a long time – but a difficult one. The good times of the 1960s were not fulfilled by the events of the 70s, 80s and 90s. Life was much harder for the Baby Boomers than they envisaged. They were living through a Youthquake that was characterised by transformations such as the gender revolution, which reshaped our views on marriage and divorce and redefined the nature of family life. Economies were restructured including a radical redistribution of work and wealth between classes; there were levels of unemployment not seen since the Great Depression, and the beginning of the information technology revolution. While Baby Boomers were living the dream of retiring at 55, changes to the retirement age in many countries, a lack of planning, and the global financial crisis has seen most continue working for longer to achieve financial security in retirement.

Youthquake for Millennials, on the other hand, will be quite different to that of their parents. The three-stage model of success being education, employment and retirement simply won't accommodate Millennials' lives, nor those of the generations that follow them as life expectancy increases. We are witnessing this with Millennials keeping their life options open, delaying the transitions to student life often taking a gap year, from student life to professional life often travelling abroad, and from single life into family life, pursuing other lifestyle choices.

This is a time of experimentation for them. Just as options have commercial value, so too do they have lifestyle values and Millennials are taking them up.

As Gratton & Scott suggest in their book *The 100-Year Life*, re-creation with new life stages requires investment in shifting identity to take on new roles, different lifestyles or the development of new skills. We are likely to shift away from age-related life stages to age-agnostic stages. This misalignment between the two models and its impact on expectations and the notion of 'success' was very well articulated by 28-year-old Sally White from Brisbane in her 2017 TEDx presentation on her personal experience of a 'Quarter-Life Crisis: Defining Millennial Success'[109.] A quarter-life crisis is defined by the University of London as 'a commitment or set of commitments within a life structure that is no longer desired, yet not perceived as a realistic target of change'.

Millennials grew up in the digital revolution that began in the 1980s, where the advancement of technology saw the shift from mechanical and analogue electronic technology to digital electronics. In that information technology environment, they became the most highly educated, diverse, media saturated and connected generation. They are now shaping the 21st century and significantly influence Australia and the greater world.

To Millennials, their voice and influence is global through the social media they fuelled and continue to fuel. It's instantly delivered to their smart phones and that's become as natural to them as the air they breathe, efficiently consumed through the artificially intelligent, personalised, platform-based, exponential models serving them. We need to embrace them, not ostracise them. We need to go beyond just listening to them as they crave to be heard. For this generation, their catch cry will be 'We're here for a good time and we're here for a long time, so we'd better take care of our world.'

Just like their parents who gave rise to the economic boom and the Americanisation of the world, this generation will give rise to the next technological boom and the Asianisation of the world.

Migration

While immigration has been the cornerstone in our rich tapestry of culture, immigration once again became a hot political and social topic in the past decade. It has again been linked to population pressures and congestion in Sydney and Melbourne and the perennial strong borders mantra that has dominated federal politics since the 2001 election.

Net overseas migration into Australia has remained above 180,000 people since 2006. After a 9.5 per cent decrease in 2017-18, migrant numbers remained relatively steady in 2018-19110.

The largest contributions to net overseas migration nationally in the year ending 30 June 2019 were the three largest states, New South Wales (36 per cent), Victoria (35 per cent) and Queensland (14 per cent).

The regions of the world where Australia's incoming migrants are born has changed significantly over time. The five years to 30 June 2019 have seen such changes, with a continued shift away from Europe and Oceania towards the regions of Asia. Migrant arrivals from South and Central Asia and North-East Asia are now higher than Oceania, with South & Central Asia clearly the highest region for immigration in 2019 (See Exhibit 5.1).

Exhibit 5.1: Overseas Migrant Arrivals — Australia — Region of Birth — year to 30 June 2014 and 2019

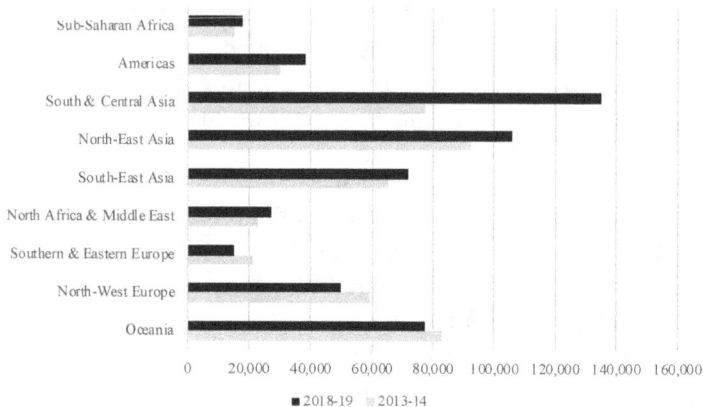

Data Source: ABS

In March 2020, Prime Minister Scott Morrison announced a new population plan that included 23,000 new visas for skilled workers who are willing to migrate to regional areas, a significant jump from 8,534 last year, along with a cut to the permanent migration program. The incentive being that if these skilled workers reside in regional cities and towns for three years, they will be eligible for permanent residency. In addition, new measures designed to attract more international students to regional universities were announced. The incentive for them is that they will be permitted to stay and work in regional Australia for 12 months after they graduate.

The policy settings that focus on skilled migration programs will form an important part of addressing the significant skills gaps present in our economy today (further discussed in chapter 7). Also, as we discussed in the last chapter with the rise of Asia, migration will form an important cultural understanding and establishing our place within it.

The demographic challenges facing the world

Over the past decade, the world's population has increased from 7 billion in 2010 to 7.7 billion in 2020. Within the next decade there are likely to be approximately 8.5 billion people. A small number of countries will account for most of the increase while others are seeing declines in their populations. At the same time, the world is growing older, faster than ever before, as life expectancy continues to rise, and the fertility rate continues to fall. This decade will bring significant demographic change, the likes of which we have not seen before.

According to the United Nations,[111] the impact of these demographic changes over the coming decade can be summarised as follows:

1. **Ageing population.** One in 11 people (19 per cent) in the world reached 65 years and over in 2019. 2018 became an inflection point whereby for the first time in history, persons aged 65 years and over, outnumbered children under five years of age.
2. **Regional growth.** Population growth rates vary across regions. Most of the predicted population growth (75 per cent) will come from developing countries in regions including Africa, Central and Southeast Asia, and Latin America.
3. **Sustainable development.** Rapid population growth presents challenges for sustainable development. These are in the areas of poverty, equality, hunger and malnutrition, quality healthcare and education.
4. **Working age population.** For some countries, growth of working-age population is creating opportunities for economic growth. Reductions in fertility have caused the working population age (25–65 years) to grow faster than other generations creating the opportunity for economic growth, but for many nations, this ratio is in decline.
5. **Life expectancy.** Life expectancy at birth has increased to a global average of 72.6 years in 2019. However, in least developed countries, this lags by 7.4 years due largely to high child and maternal mortality together with violence, conflict and disease.
6. **Declining proportion of working-age people.** The proportion of working-age people (25–64 years) relative to those over 65 years is falling around the world. Japan holds the lowest ratio in the world of 1.8. These low values reflect the potential impact on labour markets and social welfare in many countries over coming decades.
7. **Migration.** Migration has become a major component of population change in some countries. Between 2010–2020, Europe, North America, North Africa, Western Asia and Australia and New Zealand were net receivers of international migrants, while others were net senders.
8. **Australia's demographic challenges.** Let's look at the impact of some of these major demographic changes and consider how they might impact Australian society and the policy dilemmas and opportunities they present for political parties and business leaders.

First, the ageing population – the increasing proportion of older persons in a population – is poised to become one of the most significant global social transformations of the 21st century. Historically, low levels of fertility combined with increased longevity ensure that populations in virtually all countries are growing older.

According to the United Nations, between 2010 and 2030 the number of people in the world aged 65 years or over is projected to grow by 53 per cent, from 526 million to 1 billion[112] (see Exhibit 5.2). Between 2010 and 2030, Millennials (20–39 years) will remain a significant proportion (one third) of the world's population. The number of people aged 80 years or over, the 'oldest-old' persons, is growing even faster and the United Nations predicts it will almost triple between 2010 and 2030 to 434 million.

Exhibit 5.2: Annual total world population (both sexes combined) 1910–2030 (thousands)

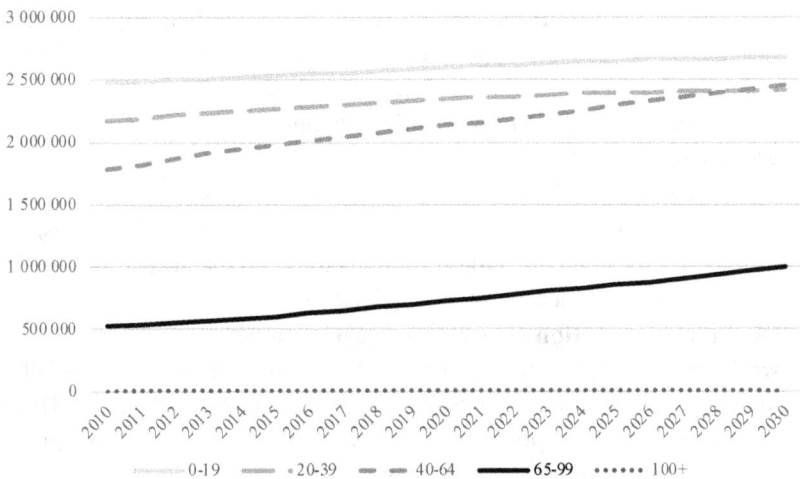

Source: United Nations, Department of Economic and Social Affairs, Population Division (2019). World Population Prospects 2019

For Australia, between 2010 and 2030 the number of people aged 65 years or over is projected to grow by 54 per cent, from 3 million to 5.4 million (see Exhibit 5.3). Between 2010 and 2030, Millennials (20–39 years) will remain a significant proportion (one third) of the world's population. The number of people aged 100+ years or over, is growing even faster and predicted to double between 2020 and 2030 to 40,000 million.

Exhibit 5.3: Annual Australian population (both sexes combined) 1910–2030 (thousands)

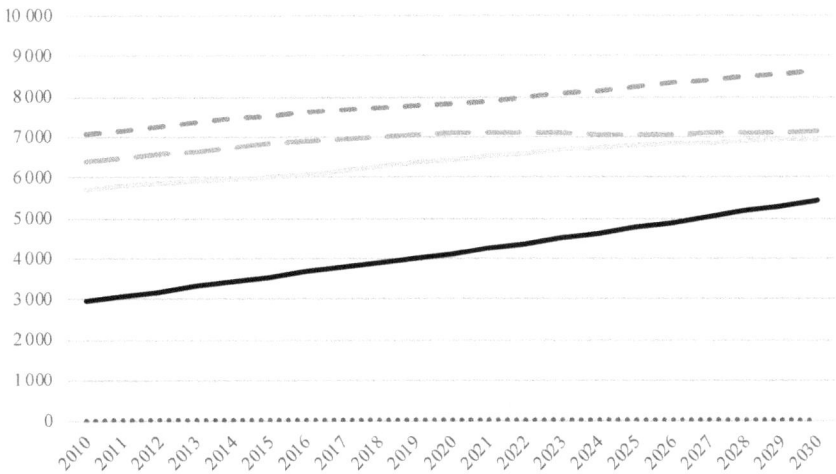

Source: United Nations, Department of Economic and Social Affairs, Population Division (2019). World Population Prospects 2019

For Australia the effects of ageing will be felt more over this coming decade than in the past due to the impact of the Baby Boomer generation retiring. This massive ageing trend has implications for nearly all aspects of society, including labour and financial markets, healthcare, the demand for goods and services such as housing, transportation and social protection, as well as family structures and their intergenerational ties.

From an economic perspective, ageing will reduce the tax revenue and add to spending pressures for the government. Over the coming decade the ageing population is projected to subtract 0.4 per cent or $20 billion from the federal government's annual growth in income in 2028-29 and add 0.3 per cent or $16 billion in spending growth[113]. In real dollar terms, this equates to an annual cost to the government's budget of around $36 billion by 2028-29. This is larger than the projected cost of Medicare that same year, the National Disability Insurance Scheme, Commonwealth funding for schools and hospitals, family tax benefit or the disability support pension.

Population ageing will have a profound effect on the potential support ratio, defined here as the number of people of working age (25 to 64 years) per person aged 65 years or over. For Australia in 2019, the ratio has 3.2 persons aged 25 to 64 for each person aged 65 or over. This is predicted to decline to 2.6 by 2030. For the Asia region, it is predicted to be 4.5, for China 3.3 and at 1.8,

Japan in 2019 had the lowest potential support ratio of all countries or areas with at least 90,000 inhabitants. Another 29 other countries or areas, mostly in Europe and the Caribbean, have potential support ratios below three (see Exhibit 5.4).

Exhibit 5.4: Estimated and projected potential support ratio, 2010 – 2100. (Persons aged 25–64 per person aged 65 or over)

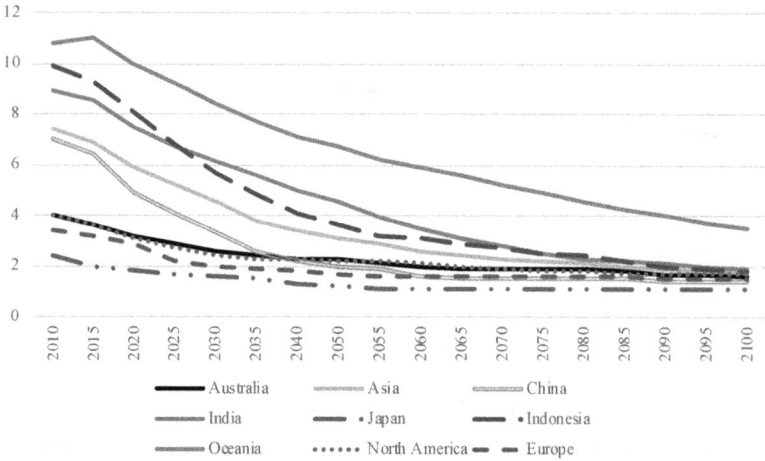

Source: United Nations, Department of Economic and Social Affairs, Population Division (2019). World Population Prospects 2019

For Australia, the old-age dependency ratio has sharply declined in the past decade from 4 to 3.2 persons aged 25 to 64 for each person aged 65 or over. This follows closely the same trend as that of the United States. This decline is predicted to continue to 2.6 by 2030. From that point, like the United States and Europe, Australia will flatten.

This means Australia must address its productivity issue detailed in chapter 2, together with the opportunities of the 4th Industrial Revolution.

Life expectancy – the 100-year life

Today life expectancy at birth exceeds 80 years on average in OECD countries (an increase of more than 10 years since 1970). Australia today has the ninth highest life expectancy at birth among OECD countries at 82.6 years[114].

The OECD reports that healthier lifestyles, higher incomes and better education have contributed to the increase in life expectancy in recent decades. A study of 35 OECD countries predicts that life expectancy gains of approximately eight years could be expected with a doubling of heath care spending, doubling income, reaching 100 per cent tertiary education and halving smoking and alcohol consumption. This illustrates the economic sustainability dilemma facing governments and citizens, as the implication would suggest a 2.5x public, private or hybrid investment requirement across these critical services to support a 100-year life expectancy.

Rising life expectancy, together with declining birth rates in most OECD countries creates an old-age dependency profile described as the number of people of retirement age as a percentage of those of employment age. Over the next 50 years, the world average is predicted to double significantly impacting the investment required by governments in the areas of health, education, employment and welfare/pensions to significantly increase relative to GDP. For example, in Australia, this increase has been estimated to be 21 per cent of GDP in 1990 increasing to 24 per cent in 2051[115]. Governments across the OECD are accordingly raising the retirement age for both women and men between 2010–2050[116]. This raises significant short and longer-term policy challenges for governments worldwide as they seek to reconcile the influential ageing population vote to that of the Millennial and now Gen Z vote.

I should note that an ageing population not only presents policy challenges for governments and businesses, but also creates enormous opportunity, as they are a valuable asset. They embody a wealth of knowledge, experience and skills that are an invaluable to society economically, socially and culturally.

Life expectancy at birth for the world reached 72.6 years in 2019, having added more than eight years since 1990. All regions shared in the rise of life expectancy over this period, but the greatest gains were in Sub-Saharan Africa, where improvements in survival have added nearly 12 years to the average length of life since 1990, reaching 61.1 years in 2019. In Central and Southern Asia, the life expectancy at birth increased by more than 11 years between 1990 and 2019, when it reached 69.9 years. Improvements in survival are expected to continue in all regions such that in 2050 the average length of life is projected to have increased to 77.1 years globally. Life expectancy at birth is highest in Australia/New Zealand, at 83.2 years in 2019, and it is expected to increase further to 87.1 in 2050.

Australia's projected life expectancy has surpassed that of other countries and regions for some time. Based on current trajectories, Australia is most likely to become the first nation to reach the average life expectancy of 100 years – post 2100 (see Exhibit 5.5).

Exhibit 5.5: Estimated and projected life expectancy at birth for both sexes (2000–2050) (Years)

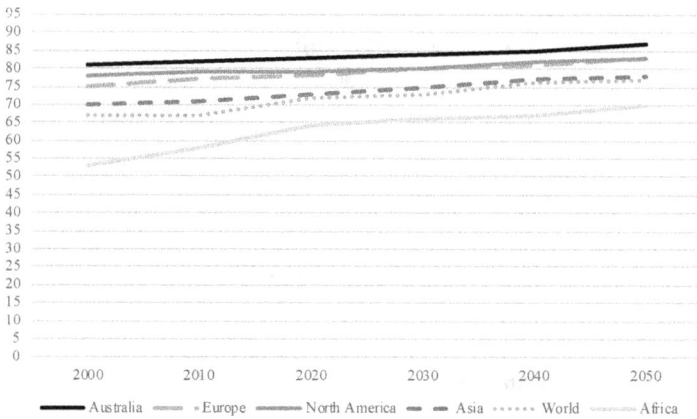

Source: United Nations, Department of Economic and Social Affairs, Population Division (2019). World Population Prospects 2019

The demographic dividend

As outlined in chapter 4, we have reached a tipping point characterised by the Asian century. So, let's consider the concentration of Millennials and the role demographic change will have. Most of the 86 per cent of Millennials in the world live in large emerging markets[117], but half are from Brazil, India, China and Indonesia[118]. These countries are also the most heavily populated. At a regional level, Asia has become the 'Millennial epicentre' where 58 per cent of the world's Millennials live.

The scale of Millennials in those countries' populations provides the opportunity for economic growth through what the International Monetary Fund describes as the demographic dividend. This is achieved in two ways. First, through investment in economic development and family welfare leading to growth in per capita income, and second through the accumulation of assets that become invested domestically, or internationally, resulting in rising national income. In developing countries such as Bangladesh, the Philippines and Vietnam this means that deploying their vast Millennial populations could lead to rapid growth projected to be 6 per cent of GDP by 2020. In contrast, in some countries where Millennial unemployment is relatively high, such as Egypt (42 per cent), Iran (29 per cent) and South Africa (53 per cent), we see lower average projected GDP growth (2–4 per cent) out to 2020 reflecting undercapitalisation on the potential demographic dividend from their vast Millennial populations.

In developed countries and regions such as the United States, the United Kingdom, Australia, New Zealand, Japan and Western Europe, lower fertility rates over many decades have led to relatively lower concentrations of Millennials in those populations. However, with ageing populations and particularly Baby Boomers at the lower end of the age bracket now reaching retirement age, Millennials in many of those developing countries are rapidly assuming greater numbers in the workforces.

As outlined earlier in this chapter, according to projections by the United Nations[119], the number of older persons globally is growing faster than for any other age group. In contrast, over the same period, the number of people 24 years old or younger will grow a mere 11 per cent and the number of people aged 25–59 will grow by 62 per cent. These projections underline the significant representation Millennials will continue to have through to 2050 as those born around 1980 reach 70 years of age. Importantly, with increased life expectancy, their democratic influence will continue to be felt.

A Millennial led Australia?

So, how do Australians feel about the influence of the Millennial demographic and their leadership over the coming decade? I put that question to the test. The results were fascinating (see Exhibit 5.6).

An overwhelming 52 per cent of Australian professionals agreed with the statement 'I believe this generation will be better transformational leaders of organisations, politics and religious institutions for this next decade, than older leaders of the past decade', compared with 20 per cent who disagreed. This result dismisses many of the stereotypes and generalisations that target this generation and provides confidence in the transformational leadership qualities this generation offers.

Now you're probably thinking of course this is what millennials would vote and they did, but it's the other generations that agreed with the statement that I believe is very important here. For example, approximately one in two Gen Xs (39–53 years) and Baby Boomers (54–72 years) agree with that statement. When it came to those who disagreed with that statement, approximately one in five Gen Xs (39–53 years) and Baby Boomers (54–72 years) disagreed with that statement. Ironically enough, even 8 per cent of Millennials (24–38 years) disagreed with that statement. Go figure?

Exhibit 5.6: Q: Millennials (24–38 years of age) have now become the largest demographic group on the planet. Statement: "I believe this generation will be better transformational leaders of organisations, politics and religious institutions for this next decade, than older leaders of the past decade" Total and by age group (%)

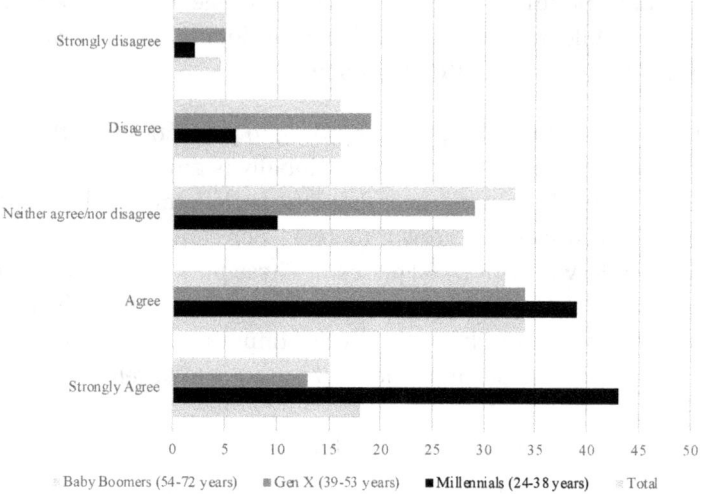

Source: Australia 2030 research Rocky Scopelliti

KEY POINTS

- Australia's population is growing by more people annually than ever before. Since 2010 the population has grown 15 per cent and is predicted to grow by 10.5 per cent this decade to 2030. The past decade has seen the urban concentration of Australian increase to 85.9 per cent of the population. We are indeed *'homo urbanis'.*

- Mental health is the top national concern for young people, with other personal concerns relating to mental health, such as coping with stress, school or study problems and body image. Our youth feel they don't have enough of a say on important public affairs issues. The growing public dialogue and experience of issues such as extreme weather events and drought are clearly affecting young people's view of the world. These results taken together clearly indicate that young people in Australia feel disenfranchised and deeply concerned and is perhaps causing them to engage in informal ways to get heard.

- The ageing population is poised to become one of the most significant global social transformations of the 21st century. Historically, low levels of fertility combined with increased longevity ensure that populations in virtually all countries are growing older. For Australia, the old-age dependency ratio has sharply declined in the past decade from 4 to 3.2 persons aged 25 to 64 for each person aged 65 or over.

- For Australia, between 2010 and 2030 the number of people aged 65 years or over is projected to grow by 54 per cent, from 3 million to 5.4 million. This means the old-age dependency ratio has sharply declined in the past decade from 4 to 3.2 persons aged 25 to 64 for each person aged 65 or over. This decline is predicted to continue to 2.6 by 2030.

- Rising life expectancy, together with declining birth rates in Australia creates an old-age dependency profile described as the number of people of retirement age as a percentage of those of employment age. Over the next 50 years, the world average is predicted to double and for Australia, this increase has been estimated to increase to 24 per cent in 2051. This raises significant short and longer-term policy challenges for the Australian government as they seek to reconcile the influential ageing population vote, to that of the Millennial vote.

- The policy settings that focus on skilled migration programs will form an important part of addressing the significant skills gaps present in our economy today. Also, with the rise of Asia, migration will form an important cultural understanding and establishing our place within it.

- Australia's projected life expectancy has surpassed that of other countries and regions for some time. Based on current trajectories, Australia is most likely to become the first nation to reach the average life expectancy of 100 years post 2100.

- An overwhelming 52 per cent of Australian professionals agreed with the statement 'I believe this generation will be better transformational leaders of organisations, politics and religious institutions for this next decade, than older leaders of the past decade', compared with 20 per cent who disagreed.

We have now crossed two major tipping points when it comes to the demographic changes within the Australian population that will set a delicate balancing act for political parties and business leaders over the coming decade.

TIPPING POINTS

1. **Youth voice matters:** Millennials and Gen Zs have now become the largest demographic groups on the planet. They are likely to be the first generations to have a 50 per cent chance of living to 100 years. This means they are also likely to see the emergence of four-generation families and so our notion of family structure will profoundly change from what we've ever known. Their proportionate representation in

society – whether as leaders in business, government or institutions be they spiritual, academic, scientific or technological – their influence will only increase from here on. Youth will shape the decade ahead and we are okay with that.

2. **Ageing population:** For Australia, the effects of ageing will be felt more over this coming decade than in the past due to the impact of the Baby Boomer generation retiring. This massive ageing trend has implications for nearly all sectors of society, including labour markets, financial markets, the demand for goods and services such as housing, transportation and social protection, as well as family structures and their intergenerational ties. The number of people aged 65 years or over is projected to grow by 54 per cent, from 3 million to 5.4 million. This means the old-age dependency ratio has sharply declined in the past decade from 4 to 3.2 persons aged 25 to 64 for each person aged 65 or over. This decline is predicted to continue to 2.6 by 2030.

Australian attitudes, beliefs and stereotypes

Throughout the decade there were many times where our generationally based attitudes, beliefs and stereotypes surfaced. Whether the context was political, business, sport, arts, music, television, news or other. This word cloud represents the major demographic changes Australia experienced during the past decade.

Of all the questions in this study, 334 Australian professionals provided commentary of a world led by Millennials. The significant agreement by all generational groups that Millennials will be better transformational leaders of organisations, politics and religious institutions for this next decade, than older leaders of the past decade was reflected in the positive sentiment (64 per cent).

I'll close this chapter by sharing some of the 334 qualitative quotes from those respondents who provided explanations for their choice.

IN AUSTRALIAN PROFESSIONALS' OWN WORDS

- *"They live in a time of heightened awareness, are more comfortable with change and social justice and it's the age of entrepreneurship – more so, they are individuals building business through solving diverse problems from social to environmental to political and technology is the vehicle by which reach and possibility has limited / no barriers to entry."*

- *"Because Millennials seem to be driven by more than the financial bottom line, bringing a people, values and ecological set of values to their leadership and work styles."*

- *"They will have the innovation and the trust of the people they lead. They will be quicker, smarter, more agile and better equipped to lead their organisation or nation so much better than we have seen before."*

- *"Absolutely – I feel that they have a bigger resonance, and are drawn to creating in the now, whilst being cognizant of the environment around them. I think instead of being self-motivated, it's more a collectivist mindset."*

- *"They bring a new style of leadership, one which embraces the qualities and values of individuals and showcases those strengths and look for better ways to strengthen their teams by bolstering numbers."*

- *"This demographic has needed to adopt to a much faster rate of change across many facets, whilst also generating disruptive approaches to solving problems. Sustained leadership capability following this approach might work in some areas but may create a void between those easily deal with change vs. those that cannot / will not."*

- *"I strongly agree that Millennials will be the generation to transform and create new ways of working to lead the five WEF areas in Social, Cultural, Economical, Technological and Political advances to lead soviet into the next decade. As a millennial myself we are a generation that likes to experiment across various industries and explore ways of working that better our social* and mental well-being rather than status. To better improve Millennials to be better leaders I believe that the previous generations such as Baby Boomers and Gen X should work collaboratively and mentor this generation around their emotional intelligence, the way they will use data more meaningfully and the repercussions of their 'I want it now attitudes, to help them manage risk either financially, emotionally, professionally and personally so they are better equipped with grit and resilience. This I feel is a missing quality to help them strive further in their pursuits upon my observations and managing the Millennial cohort both in my professional and entrepreneurial ventures".

- "As yet I see no longer term evidence that they are any better or worse leaders than their older contemporaries. Their approach is different, but we are yet to see significantly different outcomes".

- "Jacinda Ardern – enough said (yes I know technically not a "Millennial" but close enough)".

- "Last decade hasn't generated any notable leaders that created social change, not like Mandela, Hawke, Gorbachov (sic), Thatcher, Howard"

- "We are in a radical shift moving to Virtuous and Humanity cycles of business and they will be uniquely positioned to drive, lead, innovate, educate and adhere to the values and messages people (customers) seek".

- "I think we are seeing that generation standing up more than ever now and not accepting the status quo of our inadequate politician".

- "I believe their values on diversity and responsibility are stronger. Old people such as I should support them to succeed in this, not stamp out own business culture in them. They are the hope".

- "Every generation throws a hero up the pop charts" – Paul Simon

- "They are much less willing to take bullshit and are strongly driven by a desire to tear down old paradigms of working, learning and living. They are more adaptable, more agile, more "in flow" with the rapidly changing world around them and many value purpose and meaning over money and traditional institutions".

- "I believe Millennials struggle to reflect on that past and not learn from past history. I fear that Millennial leaders will focus on the flavour of the month issues as opposed to focussing on the big issues at hand. Fundamentally, leadership is about representing all people not just elevating a minority because Twitter says that's what you should do".

- "They're adaptable to change".

CHAPTER 6

Trust

"I hereby christen this budget Barbie camper
Priscilla, Queen of the desert."
(The Adventures of Priscilla, Queen of the Desert)

"Never trust a computer you can't throw out the window"

~ Steve Wozniak, co-founder Apple Inc

This word cloud represents how we feel about trust.

Source: Australia 2030 research Rocky Scopelliti

Who can forget, the Federal Liberal National Party's 2016 tedious three word election campaign slogan 'jobs and growth'. We were promised policy settings that created an unquenchable thirst for prosperity to be distributed evenly, from exuberant politicians who cracked the policy code. This confidence, however, was not matched by the electorate who had become fatigued by the political and business shenanigans throughout the past decade. But here we are in 2020 asking the same questions. Why has growth in wages, economic output and productivity reached their lowest points? Deep dives by the Reserve Bank and the Federal Treasury yield the answer: 'We don't really know'[120]. So, who do we, and what do we, trust?

In this chapter, I'll explore the interplay between how we feel about business and government and their trustworthiness.

Trust and its inequality

"The implications of the global trust crisis are deep and wide-ranging."

~ Richard Edelman, President and CEO, Edelman

Trust is the genesis of our beliefs as they relate to spirituality, society, culture, economies and now technologies. The conditions upon which we trust people, ideas and technologies – the trust trinity – have profoundly shifted away from hierarchical based vertical models, concentrated in the hands of few institutions in which we have lost faith in, in favour of a democratised horizontal based model that distributes trust among communities at global scale, real-time speed with symmetrical impact. It's what I describe as the 'renewable energy' of our digital society, culture, economies and the technologies that are intertwined in the way we live, work and play. Here's why.

According to Rachel Botsman[121], a recognised thought leader on the topic of trust, argues that institutional trust, taken on faith, kept in the hands of a few and operating behind closed doors, wasn't designed for the digital age. Botsman is right. It lacks transparency, creates transactional friction and is unfit for the 21st century, misaligned with our technological dependency and outdated for our cultural norms, with personalisation that now empowers us through our technology. Botsman proposes that we need to shift to a more contemporary model. Her point is that despite the failings of the existing trust model in its current form, it is not destroyed, but rather needs to adapt and change form to fit with our digital society. In this chapter, I'll be making the case that the time for this transition is now as it is the critical enabler to Australia's social, economic, political and technological future over the next decade.

Mind the trust inequality gap

Our views on the role of business, government and our institutions are anchored on the level of trust we have with them. Edelman began measuring trust some 20 years ago and has observed some fascinating trends reported in their 2020 Edelman Trust Barometer report122. They found that a majority of respondents in every developed market do not believe they will be better off in 2025 and that the world is diverging in two trust realities. The first they classify as the 'informed public' who are wealthier, more educated, and frequent consumers of news. They found that this group have been more trusting of every institution relative to the second group, they define as the 'mass population'. The gap between the two is referred to as the 'inequality gap'. Edelman reports that there are now eight markets (including Australia) showing all-time high inequality gaps between the two groups.

Let's explore the interplay between these dimensions and what they mean for Australia.

Australia leads the world with its trust inequality gap

As mentioned earlier, there are now eight countries showing all-time high inequality gaps between the informed public group and the mass population. At the beginning of the decade there was only one country reflecting that the number of countries with record trust inequality is at an all-time high.

For 2020, Australia held the widest gap (23), followed by France (21), Germany (20), Ireland (17), Spain (17), Italy (16), Russia (14) and Brazil (11) (see Exhibit 6.1).

Exhibit 6.1: Trust Gap – 2020 (score 1–100)

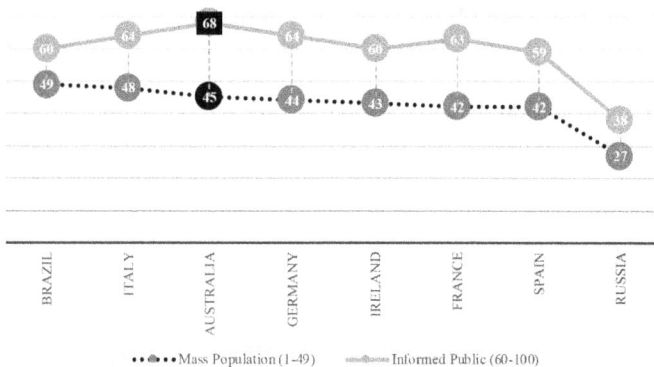

Source: Edelman

For Australia, the past decade has seen an ever widening of the trust inequality gap reaching its and the world's biggest with a 23-point gap at the end of 2020 (see Exhibit 6.2). Why is it that Australians are more polarised between the 'informed public' and the 'mass population'? Let's explore this in more detail.

Exhibit 6.2: Trust Index Australia 2012 – 2020 (1-100)

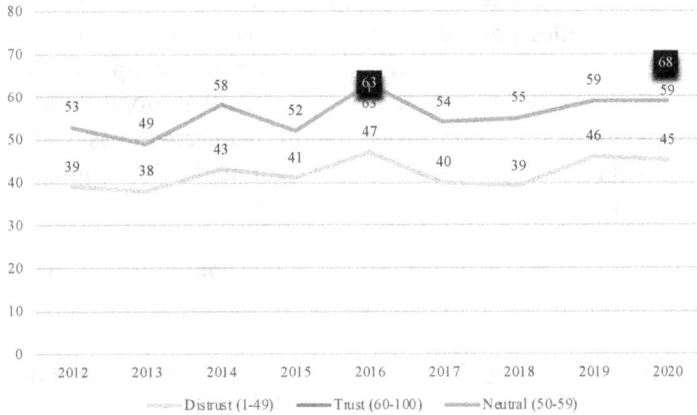

Source: Edelman

According to Edelman's research, people grant trust based on two distinct attributes: competence (defined as delivering on promises) and ethical behaviour (defined as doing the right thing and working to improve society). What they found in the 2020 study was that neither government, business, non-government organisations or the media were seen as both competent and ethical. This finding was observed at a global level and also at an Australian level.

Looking out to the next decade, the Australia 2030 research found that Australian professionals hold similar beliefs. When looking at who they trust the most and least to control their best interests, and who they expect to be the most and least ethical over the coming 10 years, we find a very similar pattern. Academic and research institutions were found to be the only organisational types to have been ranked number one for both trust and ethics attributes. In the opposite quadrant, government was found to be least trusted and ethical. Employers and NGOs are both ranked 2nd highest for ethics but low on trust as it relates to looking after people's best interest over the coming decade (see Exhibit 6.3).

While large 'public technology or scientific organisations' also had high levels of trust, they ranked equally low with the government on least ethical.

Exhibit 6.3: Trust/Ethics ranking by organisational type (most/least)

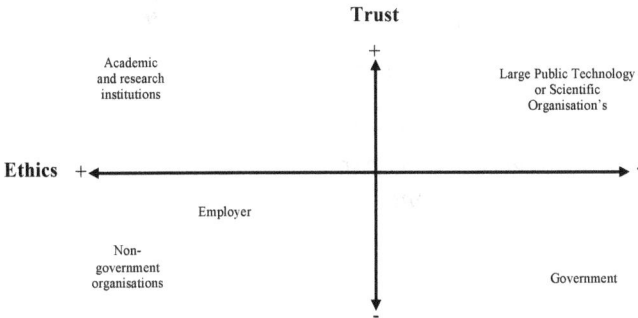

Source: Australia 2030 research Rocky Scopelliti

Distrust and its relationship to risk

"While trust is an essential ingredient in any functional relationship, be it person-to-person or brand-to-customer, it cannot shield against the corrosive effects of distrust".

"To really come to grips with this it's important to understand that distrust is not uncertainty about whether to trust, nor is it an absence of trust. It is something separate, something much darker and more damaging. The key message for business is that distrust is a major risk factor which must be monitored because it leads to customer churn, loss of market share, and a plummeting share price[123]."

~ Michelle Levine, CEO, Roy Morgan Research

There has never been a more important time in the modern world where we have required trust in institutions, organisations and the media to deal with the unprecedented impact of many global issues including COVID-19. As quickly as the virus spread around the world, so did the fake news and mixed messaging from leaders all over the world. Here is some of the most dubious and dangerous viral misinformation reported in the media:

"It's raining. We're going to get wet. And some are going to drown in the rain"

~ Jair Bolsonaro, President, Brazil

"To take Covid-19 on the chin"
[and allow the disease to spread through the population].

~ Boris Johnson, Prime Minister, United Kingdom

"I don't drink but recently I've been saying that people should not only wash their hands with vodka but also poison the virus with it – you should drink the equivalent of 40-50 millilitres of rectified spirit daily."

~ Alexander Lukashenko, Prime Minister, Belarus

" I see the disinfectant where it knocks it [Covid-19] out in a minute. One minute. And is there a way we can do something like that, by injection inside or almost a cleaning?

~ Donald Trump, President, United States of America

"I'm still going to the footy"

~ Scott Morrison, Prime Minister, Australia

Who do we believe? And what do we believe? With governments at their lowest levels of trust, for many, they were not the people we chose to listen to. Instead we relied on the people we trusted the most – our academics, scientists, researchers, experts and other institutions. The Australia 2030 research found that this reliance skyrocketed to double to 79 per cent from respondents who completed the survey during the month of March 2020, relative to respondents who completed the study over the months of January and February 2020. From its already low point, the government on the other hand declined by half to just 5 per cent during March 2020 with 'public, technology, scientific and employers remaining about the same (see Exhibit 6.4).

Exhibit 6.4: Q 'Who do you trust the most to control your best interests when it comes to the technological and scientific developments over the coming 10 years?' (%)

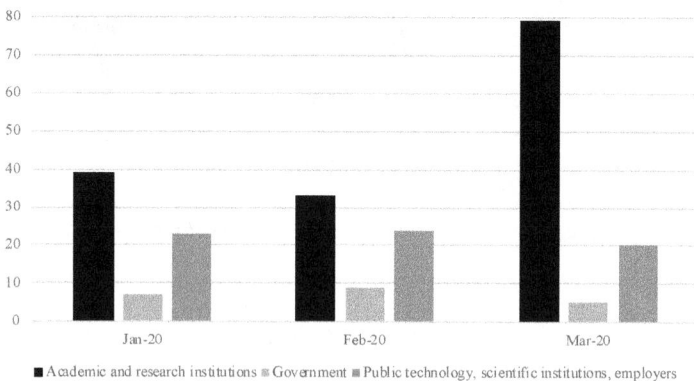

■ Academic and research institutions ▨ Government ■ Public technology, scientific institutions, employers

Source: Australia 2030 research Rocky Scopelliti

Another study by Edelman across 10 countries (Brazil, Canada, France, Germany, Italy, Japan, South Africa, South Korea, United Kingdom and the United States) between 6-10 March 2020[124] also confirmed the worldwide concerns on the questions of who and what do we trust in relation to COVID-19? For example, 74 per cent reported concerns that there was a lot of fake news and false information being spread about COVID-19 with 45 per cent reporting difficulty in finding reliable and trustworthy information about the virus and its effects. Social media has been one of the most important channels for accessing news across all demographic groups, but particularly so for Millennials (18-34 years) where 54 per cent reported getting most of their information about the virus.

A staggering 85 per cent of people reported that we needed to hear more from scientists and less from politicians. The politicisation of the crisis made people turn to social media, global health organisations such as the World Health Organisation, national health authorities, family and friends to source trustworthy information. Notably, 58 per cent of people reported that the crisis was made to seem worse for political gain with scientists (83 per cent), doctors (82 per cent) and health officials (75 per cent) emerging as the most trusted spokespeople on the virus. The three least trusted were journalists (43 per cent), countries most effected (46 per cent) and government officials (48 per cent).

Amid the concerns of access to reliable information, response to the virus, employers emerged as the most trusted to respond effectively. This included being the most trusted source of information, organisations being better prepared than their own countries and effectively and responsibly responding to the crisis. Organisations have played a critical role throughout this health and economic crisis and perhaps it is a pivotal moment in time to reclaim trust that has been eroded over the past decade.

Rethinking the relationship between trust and risk

All businesses carry risk and with that they have an uncertain life expectancy — one that been on a steady decline. The average lifespan of Fortune 500 companies decreased from 60 years in the 1920s to 15 years today. Forty per cent of the Fortune 500 companies today are predicted not to survive the next 10 years[125.]

Business have been, and should remain, drivers of innovation, creators of wealth and harbingers of economic freedom in a democratic world. The mission of a for-profit organisation is not to fulfil a philanthropic duty, nor to solely to maximise short-term shareholder value, but rather, providing the goods and services that people need or want. What has changed dramatically over time are the expectations placed on businesses. Boards of directors, management and investors of large corporations are now expected to address an array of social, economic and ecological challenges.

Business derives its social legitimacy and right to operate from the economic value it creates for society at large, from its performance for both investors and a wider network of constituencies, its partnership with governments and other agents in solving social problems, and the trust its leadership inspires in employees and society as a whole. However, as we saw in the previous section, all indicators show a sharp decline in the trust bestowed on most institutions over the past 20 years. It restores the corporation to a healthier and more relevant place in society, one originally envisaged by Adam Smith in the *Wealth of Nations,* in which he wrote that "the interests of the producer ought to be attended to only so far as it may be necessary for promoting that of the consumer."

In Smith's lifetime, shareholder capitalism occupied only a relatively small part of the way business was done. Family workshops, tradesmen, merchants, guilds, and village markets served the communities in which they lived by concentrating on end-benefits such as feeding, clothing, and providing tools of a trade. In that era, local producers were much more likely to promote the interests of consumers than to focus on disembodied shareholder profits, while large mercantile businesses worked to constrain trade and capture monopolistic rents.

More disruptive economists such as Mariana Mazzucato, an Italian-American economist once described as 'the world's scariest economist'126, are trying to change something fundamental – that is the way society thinks about economic value. She argues that the role of value creation must find its way back into its rightful place at the centre of modern economic thinking. More fulfilling jobs, less pollution, better healthcare and education and more equal pay. She challenges us with the question of what sort of economy do we want? Mazzucato argues that once that question is answered, we can then decide how to shape our economic activities.

Mazzucato has spent decades researching the economics of innovation and high-tech industry including some of the world's most innovative companies. What she uncovered was that:

"History tells us that innovation is an outcome of a massive collective effort – not just from a narrow group of young white men in California, and if we want to solve the world's biggest problems, we better understand that[127]."

Her research reveals that the development of Google's search algorithm was supported by a grant from the National Science Foundation, a US public grant-awarding body. That electric car company Tesla initially struggled to secure investment until it received a US$465 million loan from the US Department of Energy and that other companies founded by Elon Musk –Tesla, SolarCity and SpaceX – jointly benefited from nearly US$4.9 billion in public funding. The more she looked, the more she found state-funded investments including the product that symbolised Silicon Valley's engineering prowess: Apple's iPhone. Her research traced the provenance of every technology that made the iPhone. She found that the HTTP protocol, had been developed by British scientist Tim Berners-Lee. The internet began as a network of computers called Arpanet, funded by the US Department of Defense (DoD) in the 1960s. The hard disk drive, microprocessors, memory chips and LCD display were also funded by the DoD. Siri was the outcome of a Stanford Research Institute project commissioned by the Defense Advanced Research Projects Agency (DARPA) and the touchscreen was the result of graduate research at the University of Delaware, funded by the National Science Foundation and the CIA. According to Mazzucato:

"Steve Jobs has rightly been called a genius for the visionary products he conceived and marketed; this story creates a myth about the origin of Apple's success. Without the massive amount of public investment behind the computer and internet revolutions, such attributes might have led only to the invention of a new toy[128]."

The narrative of innovation by the private sector conveniently did not include that the role of the government was by design as corporations lobbied for regulatory and taxation relief. A study by Mazzucato and economist Bill Lazonick, between 2003 and 2013 of publicly listed companies in the S&P 500 Index, found that more than half of their reported earnings were used in share buybacks to boost stock prices, rather than reinvesting those earnings back into research and development. Apple started share buybacks in 2012 and by 2018 had spent nearly US$1 trillion dollars on share buybacks.

That finding presents a more fundamental problem. If it was the government, not the private sector, which had in fact assumed the risks of uncertain technological enterprises that led to the development of the 1stt, 2nd and 3rd

Industrial Revolutions, how are we going to find the next wave of technologies that characterise the 4th Industrial Revolution?

The tension of the role between trust and risk and the manifestation of distrust played out in the Australian Royal Commission into Misconduct in the Banking, Superannuation and Financial Services Industry129. Commissioner Kenneth Hayne zeroed in on the pursuit of profits and monetary gain on the part of banks and regulators that preferred negotiated outcomes over enforcement of the law in a searing indictment of the culture. In the executive Key points, Commissioner Hayne asks what the overarching motivation driving the abhorrent behaviour at the banks has been and arrives at a simple answer.

"Too often the answer seems to be greed – the pursuit of short-term profit at the expense of basic standards of honesty. How else is charging continuing advice fees to the dead to be explained,"

~ Commissioner Kenneth Hayne, Royal Commissioner into the Banking, Superannuation and Financial Services Industry

The report details that "following the Global Financial Crisis, annual profit became the defining measure of success, one that was in the interests of shareholders and therefore all Australians via super funds. But there being little threat of failure of the enterprise, and there being little competitive pressure, pursuit of profit has trumped consideration of how the profit is made".

The role of business and the government

A decade after the financial crisis, capitalism is held responsible for a variety of major societal issues, including wealth inequality and climate change. There have always been critics of capitalism, but there is now a sense that the displeasure is spreading with younger generations attracted to socialism. While capitalism has lifted hundreds of millions of people out of poverty and spurred life-changing technology, it has also been reliant on the burning of fossil fuels, and some argue that inequality isn't a faulty outcome but a fundamental feature of the system. Capitalism was supposed to make everyone better off, but long-term wage stagnation and weak productivity growth is a strong challenge to this assumption.

For example, this past decade has seen wages growth decline in Australia past that of the prior decade. A range of measures show a significant slowing in wage growth in Australia over the past five years. The Wage Price Index

(WPI) grew at an annual average of 2.2 per cent in the five years to December 2018, which compares with the average annual growth of 3.3 per cent in the previous five years to December 2013[130]. This translates into declining average weekly earnings that were frequently below 2 per cent this past decade. On some measures, it's the lowest wages growth since the Great Depression and the Second World War. A study by PricewaterhouseCoopers (PwC) of the Australian economy and business challenges for 2020 found that businesses underpay workers by around $1.35 billion each year[131].

Another major concern for Australia is the accompanying massive drop in productivity. The past three years have been the second worst period of productivity growth ever recorded, second only to the recession of the 1990s. According to PwC, a major contributor to this was Australia's lack of investment especially in research and development – the very things that should drive productivity. PwC's modelling indicates that Australia spends just 1.9 per cent of GDP on research and development and needs to spend an additional $13.7 billion to reach OECD top 10 spending levels.

The obvious problem is that no other economic model has worked any better. Mazzucato's ideas of reinventing capitalism, importantly, deviate from the 80s notion that 'value creation' is the domain of business and that government's role is to just 'fix' things.

Most Australians are concerned that the country is not investing enough in technological, scientific and skills development compared to other countries over the coming decade, and a third just don't know (see Exhibit 6.5). Modelling by PwC reveals that there is a significant economic benefit if Australia can close its digital skills gap. For example, training an additional 100,000 technology sector workers above the current levels over the next five years could add $40 billion in net present value terms to GDP over the coming 20 years[132].

Exhibit 6.5: Q: Are you concerned (or not) that Australia is (or isn't) investing enough in technological, scientific and skills development compared to other countries for the coming 10 years? (%)

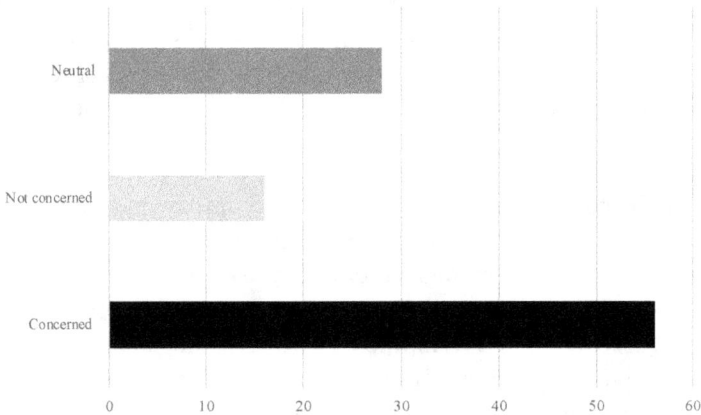

Source: Australia 2030 research Rocky Scopelliti

For Australians, the expectations on the contribution of businesses to society over the coming 10 years reflect Mazzucato's idea of the need to reinvent capitalism and the role of business in that. The notion that value creation through the allocation of capital to best projects, valued goods and services (16 per cent), together with boosting the economy (12 per cent) were the two least prioritised contributions Australian professionals felt were important for businesses to make to society over the coming decade. Corporate social responsibility emerged as the most important contribution (30 per cent), followed by improving livelihoods (23 per cent) and the creation of jobs (19 per cent). What's fascinating is that the Australia 2030 research found that those priorities fundamentally changed in March 2020 when the COVID-19 crisis hit Australia. For example, for those respondents in March 2020, boosting the economy and creating jobs became equally (43 per cent) the two most important contributions businesses could make to society over the coming decade (see Exhibit 6.6). While demographically, there was no statistical difference between gender or age in the findings, there was when it came to leaders. CEOs and board directors expressed a greater priority on corporate social responsibility than the total.

Exhibit 6.6: Q: What do you consider to be the most important contribution that businesses make to society over the coming 10 years? (%)

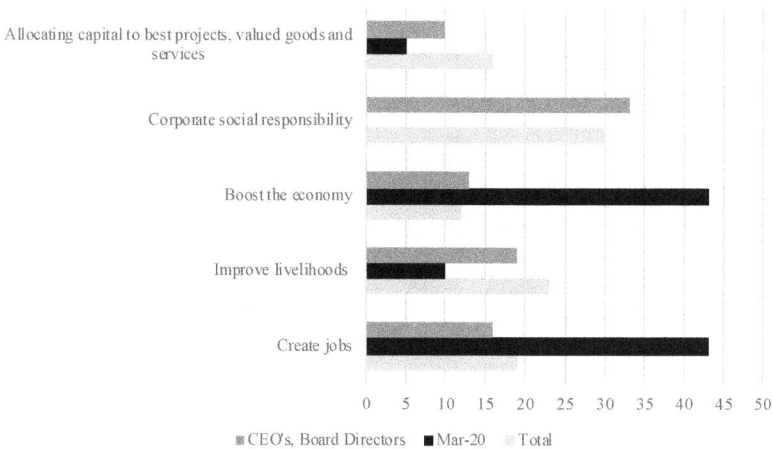

Source: Australia 2030 research Rocky Scopelliti

In democracy we trust, but not in the systems, even during the crisis

When we look at who Australians believe has the greatest role to play in making the world and Australia a better place, we find that this is not something Australians trust to institutions or government, but rather overwhelmingly, one in two see that it's individuals. This perhaps reflects the disengagement by Australian professionals and loss of trust with institutions and government over the past decade. That distrust was reinforced by respondents over March 2020 with a significant increase of 34 per cent to 85 per cent who see individuals taking responsibility and a significant corresponding decline to just 5 per cent who chose government. When we look at this demographically, we can see subtle, but significant generational preferences emerging. For example, Millennials expect employers and global/multi-national corporations to make a larger contribution and government to make a much less significant contribution relative to other demographic groups (see Exhibit 6.7).

Exhibit 6.7: Who do you believe has the greatest role to play in making the world and Australia a better place over the coming 10 years? (%)

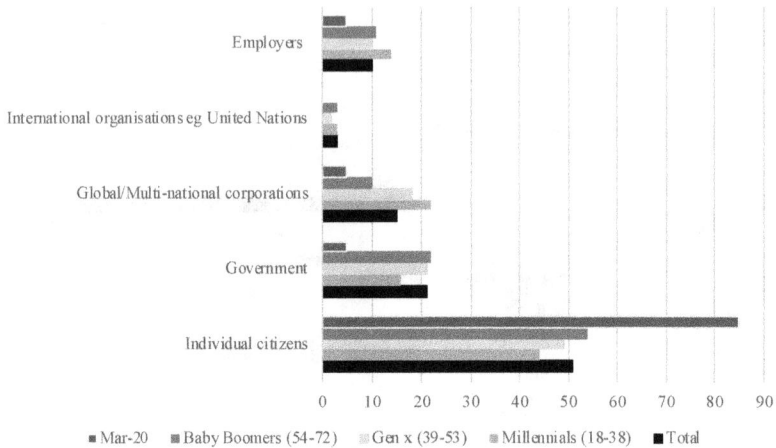

Source: Australia 2030 research Rocky Scopelliti

As discussed in this chapter, trust in our institutions, leaders and systems has fallen to an all-time low. While many people talk of recovery, I predict that we have crossed the inflection point, upon which the question has changed from 'who do we trust' to 'what do we trust'. The implications of this are profound for two key reasons. First, it suggests that restoration of trust as it was bestowed, and to recipient actors, in the past, will unlikely produce a result. Second, we will see the transition from placing our trust in governments, banks, retailers and other organisations to technology.

The Australia 2030 research found overwhelming support for that prediction with 60 per cent of respondents believing that we will trust technology more than institutions over the coming decade (see Exhibit 6.8). That skyrocketed to 90 per cent for those respondents in March 2020 with 65 per cent strongly agreeing with that statement. Demographically, Millennials held that belief more so than other age groups (76 per cent). At a professional level, 60 per cent of CEOs, directors and chairmen agreed with that belief. When it comes to those who trust robots to make decisions on our behalf, 75 per cent believe that we will trust technology more than institutions over the coming decade.

Exhibit 6.8: Q. In your opinion, do you believe we will trust technology more than institutions such as Government, banks, retailers etc over the coming 10 years? (%)

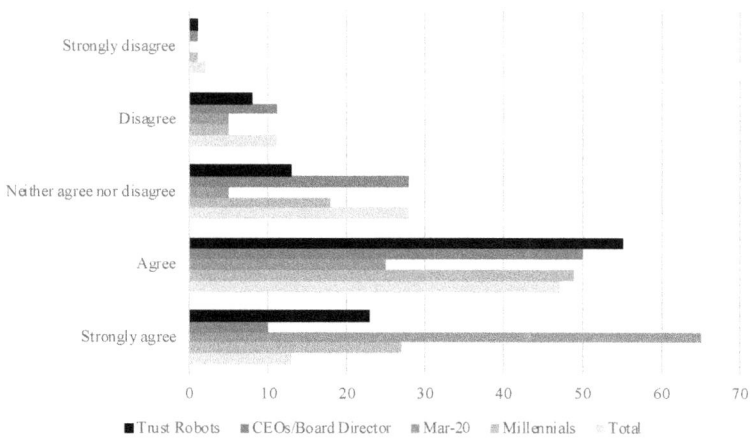

Source: Australia 2030 research Rocky Scopelliti

For Australia, like all other nations, the magnitude of the COVID-19 pandemic required government to step in to head off a health care and economic crisis. Stimulus packages, securing health care capacity, value chain security, population isolation policies and much more were invoked to protect our population. This intervention required government to play a very different role to that not seen since wartime. Importantly, that role required leadership of an electorate whose trust had been eroded. An electorate that had lost faith in its financial, spiritual and healthcare institutions. However, what has emerged is an unprecedented level of collaboration between government and business across all social, economic and political aspects of our society. That was systems leadership and the opportunity that awaits on the other side.

KEY POINTS

- Trust is the genesis of our beliefs as they relate to spirituality, society, culture, economies and technologies. The conditions upon which we trust people, ideas and platforms – the trust trinity – have profoundly shifted away from a hierarchical based vertical model, concentrated in the hands of institutions in which we have lost faith in, in favour of a democratised horizontal based model that distributes trust among communities at global scale, real-time speed with symmetrical impact. It is the 'renewable energy' of our digital society, culture, economies and the technologies that are intertwined in the way we live, work and play.

- For Australia, the past decade has seen an ever widening of the trust inequality gap reaching its and the world's biggest 23-point gap in the end of 2020. Australians are more polarised between the 'informed public' and the 'mass population' than we have ever been, and more so than any other nation.
- Looking to the next decade, the Australia 2030 research found that Australian professionals trust academic and research institutions to look after their best interests. Conversely, government was the least trusted and ethical.
- Who do we believe? And what do we believe? With government trust at its lowest levels, for many, they were not the people we chose to listen to during the COVID-19 crisis. Instead we relied on the people we trusted the most – our academics, scientists, researchers and other institutions.
- Contemporary economists now argue that the role of value creation must find its way back into its rightful place at the centre of modern economic thinking. More fulfilling jobs, less pollution, better health care and education and more equal pay. This requires a very different perspective on trust and its relationship to risk. When Australia can answer the question of what sort of economy, we want it may be that we can then decide how to shape our economic activities.
- Distrust is not uncertainty about whether to trust, nor is it an absence of trust. It is something separate, something much darker and more damaging. For business, distrust is a major risk factor that must be monitored because it is critical to its survival.
- For Australian professionals, the expectations on the contribution of businesses to society over the coming 10 years are in the areas of corporate social responsibility followed by improving livelihoods. However, those priorities fundamentally changed in March 2020 when the COVID-19 crisis hit Australia. Boosting the economy and creating jobs became the two most important contributions businesses could make to society over the coming decade.
- Australian professionals overwhelmingly believe that individuals have the greatest role to play in making the world and Australia a better place. This was reinforced by respondents during March 2020 with a significant increase of 34 per cent to 85 per cent who see individuals taking responsibility, and a significant corresponding decline to just 5 per cent who chose government.

1. **The trust inequality gap with be technologically closed.** Australian society's trust is polarised between the 'informed public', who are wealthier, more educated, and frequent consumers of news, who are more trusting of every institution relative to the 'mass population'. But for the mass population, trust in media and government has become irrevocable. For our democracy to function democratically, that gap must be minimised and ultimately closed.

2. **A new distributed trust world order emerges.** The pursuit of 'trust restoration' to hierarchical models in the hands of few is a fanciful pursuit. We have crossed the point where we hold greater trust in our own individuality that empowers us through symmetrical access to developments anywhere, anytime and anyhow.

Australian beliefs on who and what we trust

Throughout the decade there were many events where we questioned trust in our social, economic, political and technological society. This word cloud represents the decade that was trust throughout the world.

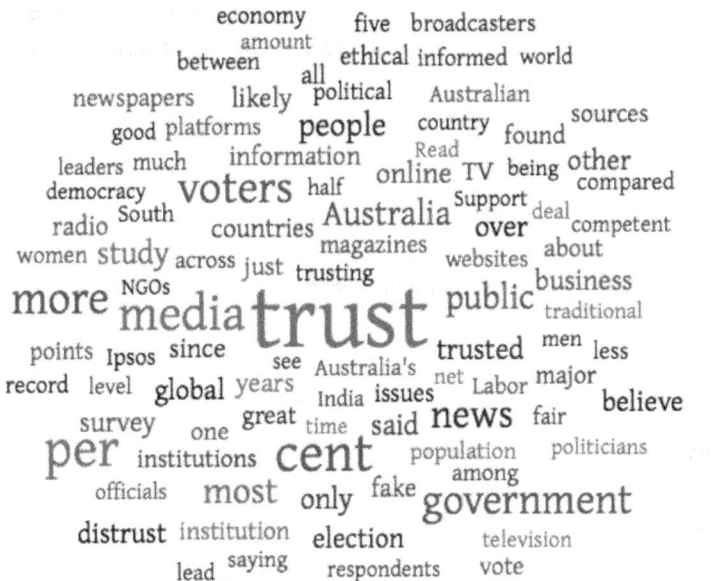

Source: Australia 2030 research Rocky Scopelliti

I'll close this chapter by sharing some of the qualitative quotes from 189 respondents who provided explanations for their choice.

IN AUSTRALIAN PROFESSIONALS' OWN WORDS

- *"Australia has lost many job creating industries and could potentially do more to drive home grown innovation which would lead to jobs".*

- *"Standard of living will be put at risk as we slowly slip into menial 'service' jobs run by multinational companies that provide no infrastructure or investment back into the community. GIG Economy is a fast race to the bottom of casualisation of work and reverse bidding for an hourly rate".*

- *"Because I see little to no long-term insight or commitment from the Government that investing in technological, scientific and skills development will pay off in the future. It's too focused on the next election and not the next generation".*

- *I am concerned as pretty much right now we are a dumb and rich economy, as we pretty much dig things up and export, we don't value add, and don't export much in technology to the world".*

- *"We need to be investing for the "digital" age and the shifts required to combat climate change".*

- *"Australia, as someone put it, is "lucky but dumb". Investment into anything seen as intellectual is frowned upon by the Government, as they actively turn their back on science.*

- *"Australia is viewed favourably around the world and has held their own for an extended period of time, and I expect the same going forward".*

- *"I am concerned that Australia is not investing enough to keep up with the global race. e.g. Dubai is really small city, but they are already looking through the future and heavily investing into the technology, science and infrastructure".*

- *"High cost of education and reduction in Education exports. We need to lower the barriers to education and the culture surrounding it – we need to be constantly learning to adapt, reskill and maintain a competitive advantage".*

- *"Investment in research and scientific skills is critical over a 10-year period as we can select the best innovations that will address climate change, inequality, treatment of animals and the wider environment, together with evolving new jobs as others disappear through irrelevance or disruption".*

- *"We aren't creating the curious child. We need to invest in developing both technology skills as well as scientific developments".*

- *"We are ranked 22 on the innovation scale. A country of our wealth with access to knowledge and research should be performing much higher on this scale. Our geographical location is not going to change so we should be working three times as hard to innovate and be world leaders in many fields. If we do not do this our competitiveness will continue to decline and our emerging neighbours in the Asia pacific region will surpass us and we will be known for a county that supplies clean air (when there are no bushfires) so we will become a residential suburb of the Asian region."*

CHAPTER 7

Ingenuity

"Just because people have Neolithic tools, Inspector, doesn't mean they have Neolithic minds." (Rabbit-Proof Fence)

"The best way to foster a more cohesive and inclusive society is to provide everybody with a decent job and income. Here in Davos, we are creating a public-private platform to give one billion people the skills they need in the age of the Fourth Industrial Revolution. The scale and urgency of this transformation calls for nothing short of a reskilling revolution[133]."

~ Klaus Schwab, founder and executive chairman,
World Economic Forum.

This cloud depicts how we feel about knowledge, skills and the way we learn.

Industrial transformations traditional organisation's value impact **Digital** collaboration Innovation capacity business intelligence **Agile** technology COVID strategies Revolution emerging **Data** organisations crisis **failure** respondents workforces technological more Purpose industries transformation between leaders **Robotics** technologies Australian changing research Millennials juvenescence future adapt human ability economy **Gap** artificial thinking biggest **AI** management found executives businesses change Internet **Culture** workforce employees Forum models decade speed global enterprises Australia skills economic scale workplace Digtal balance performance **Experimentation Reskilling** opportunities demographic leadership model world prioritise jobs work organisational people learning years professionals **Automation Ingenuity** Enterprise survival **Inclusion** systems discrimination processes autonomous Experiementation **Industry** financial new Australia's

Source: Australia 2030 research Rocky Scopelliti

A reskilling revolution was launched at this year's 2020 World Economic Forum in Davos, Switzerland. This multi stakeholder initiative has the bold aim to provide better education, new skills and better work to a billion people around the world by 2030. The objective is to prepare the global workforce with the skills needed to future-proof their careers against the expected displacement of millions of jobs and skills instability as a result of technological change. For Australia, our economy is in transition and the capacity to transform our skills will be a critical success factor for a prosperous, equitable, inclusive and fair society.

We know the digital era has created this disruption and most industries are struggling to keep pace with the demands for new skills. As leaders of enterprise, government and academia, it is our joint responsibility, indeed our obligation, to prepare our existing workforce and those entering the workforce for a very different future.

Innovations in the emerging technologies of the 4th Industrial Revolution certainly have the potential to be a power for good – connecting billions of people globally, solving complex scientific and medical challenges and addressing world poverty. However, governments, policy makers, education and industry are struggling to keep up with the rapid changes, be it legally, ethically, culturally, socially or economically.

How we think about what we once believed 'normal', its 'rules' and what we're influenced by is all changing. This is being driven by demand for greater access to information anywhere, anytime and on any device. The power is now in the hands of every digitally connected citizen, personalised and democratised. Artificial intelligence (AI) and machine learning (ML) will become some of the biggest catalysts of change to the human workforce. Jobs are changing, that's irrefutable.

So, what do these jobs look like? What will be automated? What processes can benefit from harnessing 4th Industrial Revolution technologies? What blend of technical, hard and soft skills will be required? These are all critical questions Australians should be addressing.

We need roles people can step into that will fuel our economy and launch Australia into a far more prosperous and sustainable future. Jobs like cyber specialists, engineers, data analysts, programmers, robotic repairs, AI, ML and Internet of Things integrators and scientists. Our ability to handle unpredictable situations that require out-of-the-box thinking – critical thinking, problem solving, empathy, understanding, creativity and collaboration – are all skills that must be encouraged and developed. These are the lifelong transferable skills irrespective of the career choices we make.

Think of the rapid advances in science and technology changing health care and medicine. COVID-19 brought this to our attention on a global scale with the speed at which the world's scientists are searching for effective vaccines and treatments. How we stay healthy in 2030 will be very different from today. Remote health monitoring, robotic surgery, the 3D printing of organs and bones are now realities. This all creates the need for workforces that can innovate, build, operate, fix and feed the machines.

For this chapter, I'll draw upon the Enterprise 4.0 research I had the privilege to lead in Optus Business, in the capacity of Director, Centre for Industry 4.0 and published in a series of reports[134]. That research, the first and largest study of its type in Australia, sort to understand how ready Australian organisations and their workforces are for the 4thIndustrial Revolution.

The importance of skills and culture in the 4th Industrial Revolution.

"We are at the beginning of a revolution that is fundamentally changing the way we live, work and relate to one another".

"Businesses, industries and corporations will face continuous Darwinian pressure and as such, the philosophy of "always in beta" always evolving will become more prevalent"[135].

~ Klaus Schwab, founder and executive chairman, World Economic Forum

This philosophy by Schwab directly informs the question at hand–in this environment, how do we increase our capacity to adapt? The evidence to date suggests that nations, industries and corporations are yet to fully capitalise on the benefits of the current digital revolution which may well be the single biggest barrier to unlocking the potential of the 4th Industrial Revolution. This revolution is characterised by emerging technology breakthroughs with potentially highly disruptive effects in the areas of artificial intelligence, robotics, the Internet of Things (IoT), autonomous vehicles, 3D printing, biotechnology, nanotechnology, materials science, energy storage, blockchain, quantum computing, 5th generation mobile networks and many more.

Today, leaders are aware of the need to upskill, reskill and recruit employees with the required digital capabilities for this new way of work. The World Economic Forum's *The Reskilling Revolution 2020* report states that 75 million jobs are expected to be displaced due to automation and technological integration in the coming years, however, this transformation will also create demand for an estimated 133 million new jobs[136]. This optimism is certainly reflected in the Australia 2030 research, which found that Australian professionals across all demographic profiles believe that technology will create more jobs than it destroys over the coming 10 years. Particularly so for Millennials (63 per cent) (see Exhibit 7.1).

Exhibit 7.1: Q. In your opinion, do you believe technology will create more jobs than it destroys over the coming 10 years? (%)

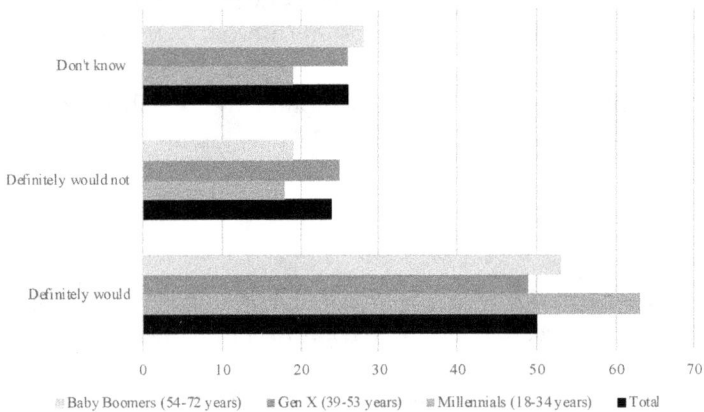

Source: Australia 2030 research Rocky Scopelliti

The most in-demand skills will include both 'hard' and 'soft' skills (see Exhibit 7.2) such as analytical thinking, innovation, active learning and broad learning strategies. They will be required for roles ranging from data analysts and scientists through to software engineers and application developers[137].

Exhibit 7.2: Emerging roles

Emerging Roles	Growth Roles	New Specialist Roles
• Data analysts and scientists	• Customer service workers	• Artificial intelligence and machine learning
• Software and applications Developers	• Sales and marketing professionals	
		• Big data
• Ecommerce	• Training and development	
		• Process automation experts
• Social media specialists, roles that are significantly based on and enhanced using technology	• People and culture	• Information security analysts
	• Organisational development specialists	• User experience and human-machine interaction designers
	• Innovation managers	• Robotics engineers

Source: World Economic Forum, (2018) Future of Jobs

As with all past industrial revolutions, human ingenuity can explain the developments that have led our evolution to this point. Our resourcefulness transcended global boundaries that fundamentally changed our labour, production and financial markets. But the speed, scale and impact of this 4th Industrial Revolution presents us with a very different set of conditions with the significant challenge – how do we close the ingenuity gap?

Australia's ingenuity gap

The ingenuity gap was defined by Thomas Homer-Dixon as the shortfall between the rapidly rising need for ingenuity and its adequate supply[138]. The unpredictability of this 4th Industrial Revolution's speed, scale and impact, driven by the associated emerging technologies, means we can't always supply the ingenuity we need at the right times, in the right places and in the right quantities. For example, it's been reported that Australia will need up to 161,000 artificial intelligence specialists by 2030[139] and other reports predict technology skills shortfalls of 100,000 by 2024[140].

Without effective strategies developed and planned collaboratively by enterprise, academia and government, Australia risks widening its ingenuity gap. The situation means we must evolve how we develop and utilise human capital.

The urgency with which Australia's well documented ICT skills shortage must be addressed is heightened by findings that outline the technologies enterprises expect will have the biggest impact on their business. The *Enterprise 4.0* study questioned Australian executives to understand which emerging technologies they felt would have the greatest impact on innovation and/or disruption in their industry over the next three years (see Exhibit 7.3).

A notable 85 per cent of respondents place cyber/information security disruption on their three-year horizon. For more than half of the Australian executives and professionals surveyed, five other 4th Industrial Revolution emerging technologies stood out:

- Big data, analytics & algorithms (81 per cent)
- Application programming interface (76 per cent)
- Artificial intelligence (74 per cent)
- The Internet of Things (55 per cent)
- Advanced 5G wireless networks (44 per cent).

Cyber security, artificial intelligence (AI), the Internet of Things (IoT), application programming interfaces (APIs), and big data technologies are expected to drive the biggest changes during the next three years. Recruiting and reskilling employees in these areas needs to be an immediate focus to unlock the value they bring.

Exhibit 7.3: What impact do you believe each of the following technologies will have on innovation and/or disruption in your industry in the next three years? (%)

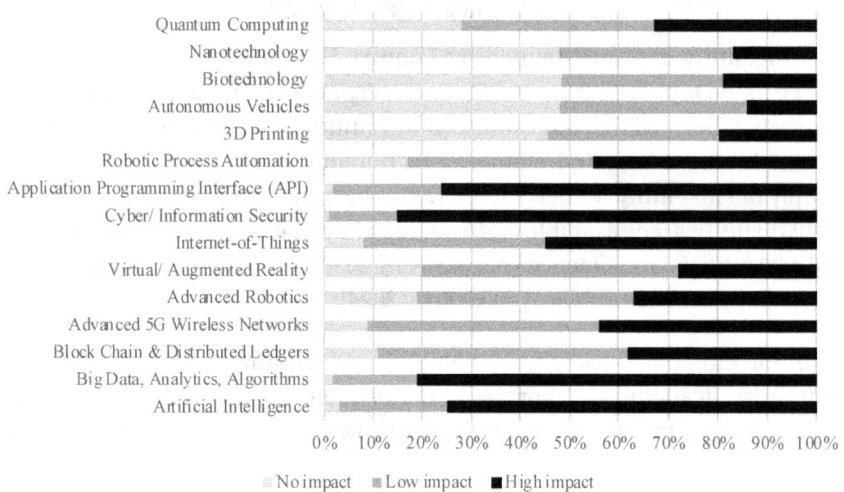

Source: Optus Business' Centre for Industry 4.0

Purpose, inclusion and culture

Australian enterprises must embrace experimentation in product development and decision-making to create the innovative digital culture needed in the 4th Industrial Revolution. It will also help to attract and retain workers with digital skills in a cost-effective way.

The *Enterprise 4.0* research found 54 per cent of Australian leaders reported that their enterprises do not tolerate innovation failure nor encourage risk taking. Another 38 per cent said these practices are a 'career limiting move' and 'encouraged, but in name only'. So, if half of our leaders hold that belief, then what does our broader workforce culture believe? For some reason, experimentation and innovation failure have become punishable, rather than being encouraged as an integral part of the learning process. This demonstrates the cultural shift needed to accelerate innovation and ingenuity among Australian enterprises.

We also need to change the models we use. Despite the need to constantly evolve and refine offerings, one third of enterprises continue to use traditional and static models to optimise business processes. This results in enterprises not unlocking the full value of the digital technologies in which they are investing. Nor are they providing an environment whereby employees can reskill to deliver value back to the business.

Enterprises united by a transformational purpose that goes beyond a mission statement are more likely to achieve exponential growth. The *Enterprise 4.0* research found that only 25 per cent of Australian enterprises have successfully fostered this culture. It means businesses need to find a way to unite their workforce behind a shared goal that makes a difference to more than financial performance.

While greater autonomy can lead to better performance in the 4th Industrial Revolution, decision-making is often still reserved for senior leaders in Australia. Forty-two per cent of enterprises in *Enterprise 4.0* research reported that they use traditional, top-down command and control processes. This practice can slow the speed at which decisions are made and acted upon.

Empowering more employees to make more meaningful decisions that have an impact on performance will accelerate the rate at which Australian enterprises can evolve structures for the 4th Industrial Revolution. The most successful organisations in the 4th Industrial Revolution will leverage small, multi-disciplinary, networked and self-organising teams. Currently,

30 per cent of Australian enterprises are designed this way, with many more beginning to lay the foundations for this type of structure (see Exhibit 7.4).

Exhibit 7.4: Q. Does your organisation operate with large, hierarchical structures or sall, multi-disciplinary, self-organising teams? (%)

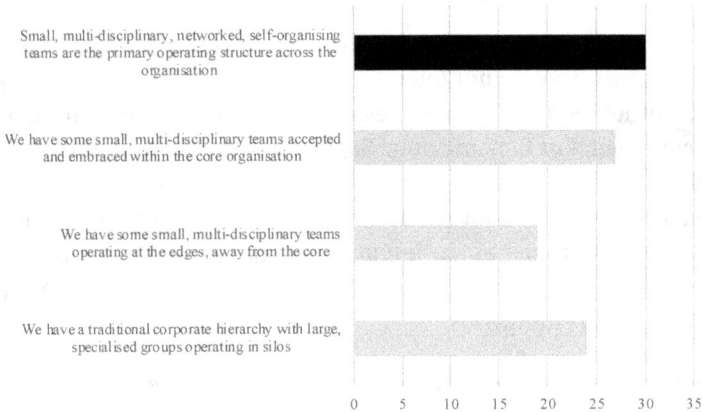

Source: Optus Business' Centre for Industry 4.0

Australia's ingenuity gap is a leadership issue, not a technological issue

So, how do Australian CEOs and their boards rate their organisation's readiness for the 4th Industrial Revolution, compared to executives that report them? The *Enterprise 4.0* research revealed there are significant gaps in how CEOs and board members rate their enterprise's agility compared with other C-level management roles. That leadership gap was tested through the COVID-19 crisis that saw organisations struggling to adapt in an agile manner, their technology and operations, workforces, supply chains and in some cases, their business models, around the restrictions being applied around the world.

The *Enterprise 4.0* research found the biggest gaps between leaders and management exist in the cultural and operational characteristics associated with businesses that have experienced exponential growth. Culture was found to have had the largest gap whereby CEOs and board members were found to have had twice the agility perception of their organisation's culture, relative to management. This discord exists at a capability, cultural and operational level in how both groups view their ability to leverage 4th Industrial Revolution technologies (see Exhibit 7.5).

Exhibit 7.5: 4th Industrial Revolution readiness: CEOs + Board Members versus C-Level Management (%)

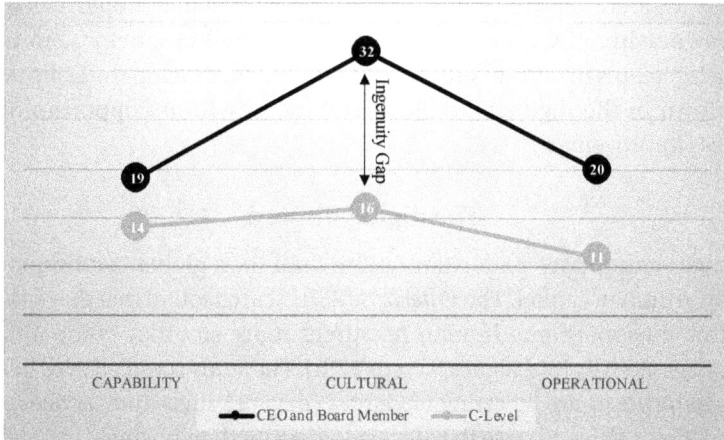

Source: Optus Business Centre for Industry 4.0

The speed at which enterprises must communicate, collaborate and adapt in the 4th Industrial Revolution means leaders cannot afford to have differing views on what can and can't be achieved. Over estimation can lead to inadequate business continuity when such events occur as evidenced through the COVID-19 crisis.

This belief among CEOs and board members that their organisations are more digitally enabled and culturally transformed than they actually are could hinder their ability to adapt. The misalignment could mean traditional processes are maintained, capacity to respond to a crisis is reduced, opportunities to innovate are missed and access to industry talent is limited.

Deloitte's *The Industry 4.0 Paradox* research demonstrates the impact this discord creates. Supply chain processes are often rated as having the most to gain by 4th Industrial Revolution influences through technologies such as advanced robotics, automation and autonomous vehicles. However, those responsible for managing those departments rarely have a seat at the management table. It means decisions and investments are not always tailored to current and future requirements.[141]

Aligning beliefs with an informed reality – based on information/data – will ensure performance metrics are realistic, training programs are tailored for current needs, industry partnerships are leveraged, and investments are focused in the right areas.

Having the right workforce skills and corporate culture will enable the broader business to operate in the way the CEO and board members expect, which today is inflated relative to management. It will enable employees to take more ownership of their decisions and responsibilities, help them inform leadership teams of current business performance, develop the skills needed to perform in the digital workplace and identify future opportunities for business improvement.

The digital discord

Australian enterprises understand more than their global counterparts the priority of human capital. The *Enterprise 4.0* research found that they prioritise talent management and human resources more so today compared with enterprises globally (3rd highest versus 12th). The study revealed gaps between where enterprises are investing resources (new products and services), what they believe is going to have the biggest impact on their business (models and economic landscapes) and what will have the greatest impact (labour).

The findings reveal Australian enterprises see the emergence of new delivery models and changing regulatory environments as having the biggest business impact over the coming five years. This heightens the urgency at which enterprises in Australia must respond to remain globally competitive.

Despite this, the emergence of new business and delivery models was considered to have less impact over the next five years by Australian enterprises relative to their global counterparts. At a closer look, however, we can see Australian executives rank the emergence of more powerful and tech savvy customers (40 per cent) and smart autonomous technologies (40 per cent) as having the greatest impact over the coming five years; this is considerably higher than their global counterparts (30 per cent and 31 per cent respectively).

People are the most crucial element in any successful business transformation. This is because the technologies are only as good as the people operating them. It's also why enterprises must prioritise developing digital skills and facilitating digital cultures.

Enterprises can augment their transformation with effective digital collaboration technologies. These act as a catalyst for a plethora of positive changes by enabling enterprises to:

1. **Accelerate and decentralise decision-making.** Employees can securely share assets or intellectual property inside and remotely from the

enterprise. This means issues are easily resolved and that data insights are acted upon in near real-time so that opportunities are realised. An example of this exposure was the security risks that manifested to organisations during COVID-19 with offshore centralised operations throughout Asia that could not operate remotely when workplace isolation measures were enforced.

2. **Bring the business closer together digitally.** Mandating the use of collaboration and knowledge-sharing platforms encourages a growth-centred mindset. Inspired by their peers, employees will be motivated to drive change and acquire new skills. It is also a forum for providing feedback on current abilities. An example of the importance of this was the inadequate provision of communication and collaboration technologies and training deployed in workforces' standard operating environments to enable them to work remotely during the COVID-19 crisis.

3. **Encourage new ways of working.** The significant degree of change delivered through improved collaboration and knowledge sharing inspires enterprises to see what else they can achieve with technology. It is an entry-point for further technological change and a way to re-skill employees in a streamlined way. For example, the COVID-19 crisis has provided many leaders with the opportunity to rethink the future of work that centres on outcomes that are delivered in a distributed model, and the benefits that it provides opposed to the capital-intensive centralised models.

These capabilities help enterprises to bridge the ingenuity gap between digital technology adoption, skills and culture so that organisational demands are met in an adaptable and agile way. The question for today's leaders as we restart the economy, and 'come out from underneath the doona' will be, does the organisation revert to the past, or will it reprioritise its initiatives within its strategic plans for this decade to adapt and progress.

The national systems leadership innovation imperative

In chapter 6, we highlighted the significant gaps by Australian professionals on questions of who they trust most and least and who they perceive to be most and least ethical when it comes to looking after their best interests over the coming decade. The results suggest a very low level of alignment between the interests of industry, government and our institutions. However, partnerships between industry, government and academia are essential to success in the 4th Industrial Revolution. The speed and scale of change means no organisation can be expected to have all the answers or skills in-house. We all experienced this through the COVID-19 crisis by observing the unprecedented collaboration and partnerships between industries, governments, regulators, academia and especially with our scientific and technological institutions in terms of decision making, implementation and enforcement. That's why Australia has come through this crisis better than most other nations. In other words, Australia's ability to harness and augment its resources, be they intellectual, cultural, biological, environmental, digital or physical, through systems leadership demonstrates our nation's potential to be a leader in the 4th Industrial Revolution.

Systems leadership can be achieved by strengthening ties with enterprises focused on developing digital skills and digital cultures, joining forces with universities to align curriculums with digital needs, and participating in accelerator and innovation-focused programs. This provides a forum to share learnings, ensures workforces are equipped with the skills they need and helps enterprises forecast future requirements.

Despite the obvious benefits of partnering and collaboration, Australia's rate of collaboration between industry and researchers (at 2–3%) is the lowest in the OECD. Australian businesses do not have as much internal research expertise as key comparator countries either (see Exhibit 7.6). At 43 per cent, Australia's proportion of researchers employed in business is significantly lower than countries such as Germany (56%), South Korea (79%) and Israel (84%).

Exhibit 7.6. Australia has the lowest collaboration rate between organisations and researchers (%)

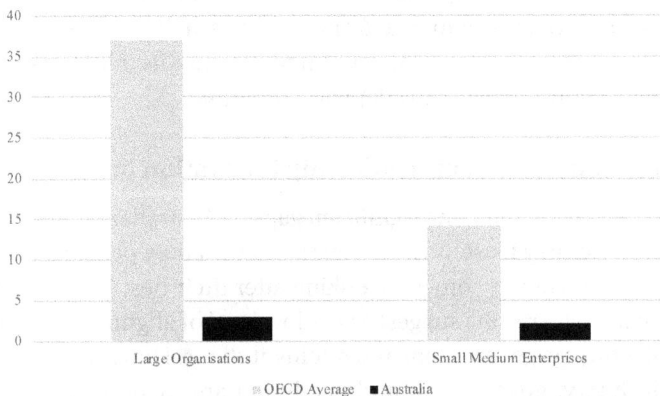

Source: Australian Department of Industry, Science, Energy and Resources

Australian enterprises must overcome their aversion to innovation failure to encourage more digital innovation, flatten hierarchical structures and decentralise decision-making in a collaborative manner. This can be achieved by adopting agile innovation methodologies to optimise processes through experimentation, but that prioritise human capital in every aspect.

People are the most crucial element in any successful nation, industry or business transformation. This is because the technologies are only as good as the people operating them. It's also why enterprises must prioritise developing digital and data skills and facilitating digital cultures. So how do Australian professionals think about their skills, their workforces and the coming decade?

Work-life balance and flexibility

The Australia 2030 research identified that overall 'work life balance and flexibility' was the biggest concern about job prospects (38 per cent) over the coming decade and this was consistent across all demographic groups (see Exhibit 7.7).

Australians are reported to have one of the worst levels of work-life balance among other OECD countries[142]. Relative to many other nations, Australia ranks in the bottom third of OECD countries when it comes to working long hours and time devoted to leisure and personal care. This was evident in the Australia 2030 research by how easily its importance declined with more than half of those respondents in March 2020 reprioritising 'reskilling and support programs' (50 per cent) and 'not enough jobs' (30 per cent) as their major concerns about their job prospects over the coming decade.

Exhibit 7.7: Q. What are your biggest concerns about your job prospects over the coming decade? (%)

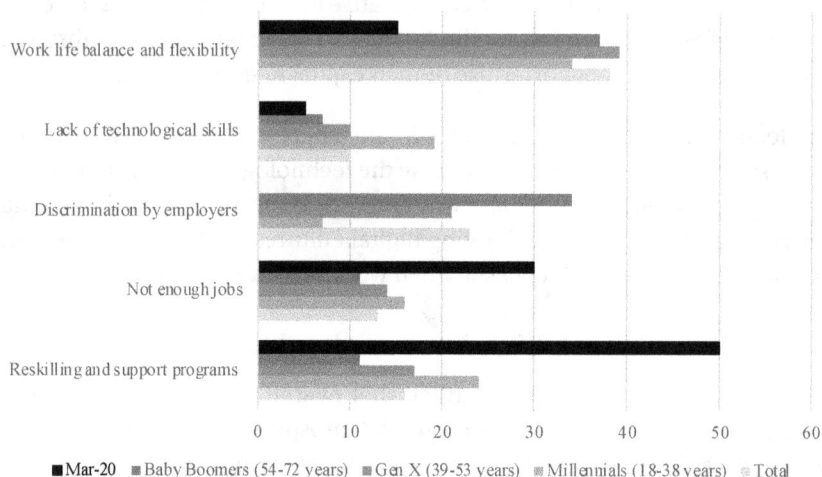

■ Mar-20   ■ Baby Boomers (54-72 years)   ■ Gen X (39-53 years)   ■ Millennials (18-38 years)   ■ Total

Source: Australia 2030 research Rocky Scopelliti

Interestingly, 24 per cent of Millennials (18-38 years) selected 'reskilling and support programs', significantly more than other demographic groups. This reflects a more commonly known aspect of employment retention with this demographic associated with personal development. Counter intuitively though, 19 per cent of Millennials are concerned with 'lack of technological skills' which may reflect a higher expectation by them on technological advances and its role in the workplace. Interestingly, and again counter intuitively, Gen Xs and Baby Boomers were least concerned about the 'lack of technological skills' which may reflect their adaption of the use of technology in the workplace longer than Millennials. Personal growth/career advancement was found to be three times more important for Millennials than other demographic groups and is the most important attribute for them when considering future opportunities and roles.

Compared to other demographic groups Millennials were the least concerned about 'discrimination by employers' but it was a concern for 34 per cent of Baby Boomers (54-72 years), and 21 per cent of Gen Xs (39-53 years). According to a study by Robert Walters, discrimination based on gender, race, disability and age in the workplace was reportedly widespread in Australia with 82 per cent of respondents (15.5 million professionals), acknowledging it in today's workforce [143], particularly women, where 9 in 10 felt it existed in the workplace. Each year the Australian Human Rights Commission (AHRC) receives more complaints about disability discrimination (42 per cent) than any other form of discrimination. This is followed by sex discrimination (27 per cent) and racial discrimination (14 per cent).

When it comes to what will influence Australian professionals over the coming decade when considering jobs and opportunities, its clearly about seeking a sense of purpose, inclusion and making an impact on society in their work. This was found across all demographic groups and particularly so for Baby Boomers who rank it number one (see Exhibit 7.8). The importance of purpose to employees and employers in the workplace is widely reported, but what does this really mean?

Exhibit 7.8: Q. In your opinion, what will be the most important criteria when you consider job or work opportunities over the coming 10 years? (%)

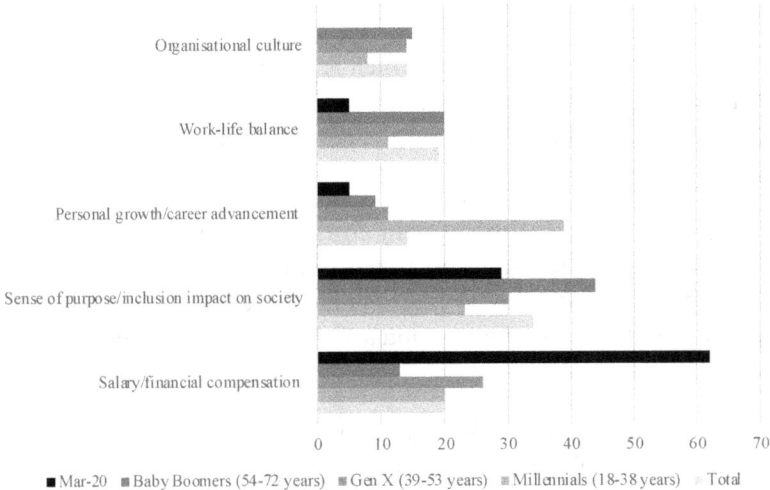

Source: Australia 2030 research Rocky Scopelliti

The priority to address work-life balance has seen that attribute become the third most important consideration for Australian professionals when thinking about job opportunities over the next 10 years. Again, we see how easily we reprioritise organisational culture in favour of salary and financial compensation during a crisis. The Australia 2030 research found that salary and financial compensation skyrocketed by 42 per cent of the total to 62 per cent for those respondents in March 2020, which became the eye of the COVID-19 storm.

There are many opportunities for Australian leaders to revisit both work-life balance and flexibility post COVID-19 when these concerns were tested. For those of us who had to juggle home schooling, multiple working parents in the one household and isolation, this was quite challenging. However, what we have learned is that for many, both these can coexist and perhaps should not be traded off against salary and financial compensation in ordinary times.

It's time for Australian leaders to get their 'juvenescence' on

As discussed, the 4th Industrial Revolution differs from the past three industrial revolutions with its speed, scale and impact. It will be a cyber-physical system characterised by new technologies that are augmenting our digital, physical, environmental and biological worlds. A key argument I presented in my last book *Youthquake 4.0* was that organisational survival is no longer a function of a one-off large-scale transformation event, but rather an organisation's capacity to continuously adapt to the fast-changing environment around them.

I described this as 'juvenescence', which is defined as the constant state of youthfulness. Juvenescence is the secret behind why some organisations survive and others don't. Put simply, it's their capacity to adapt to the environment around them that explains their survival.

We are living in the age of accelerated transformational change. While transformation itself is not necessarily new to us, the frequency, pace and impact of transformation today is. There has never been a more important time to stop and consider the question *will digital transformations guarantee organisational survival?*

Let's take a look at how likely Australian professionals believe that their organisations will still be around in the next 10 years. The average across the three-month Australia 2030 research suggests that Australian professionals were very confident of their organisation's survival with 69 per cent of respondents indicating 'likely' and only 8 per cent indicating 'unlikely' and 23 per cent 'unknown'. However, that confidence fell away dramatically throughout March 2020 as the unprecedented economic impact of COVID-19 hit industries such as airlines, retail, hospitality and their workforces and workplaces. In March 2020 only 23 per cent of respondents indicated it was 'likely' that their organisations would be around in 10 years (a 3x drop) and a staggering 48 per cent indicating 'unlikely' (a 6x increase) and 29 per cent 'unknown' (see Exhibit 7.9).

Exhibit 7.9: Q How likely do you think your company will be around in 10 years? (%)

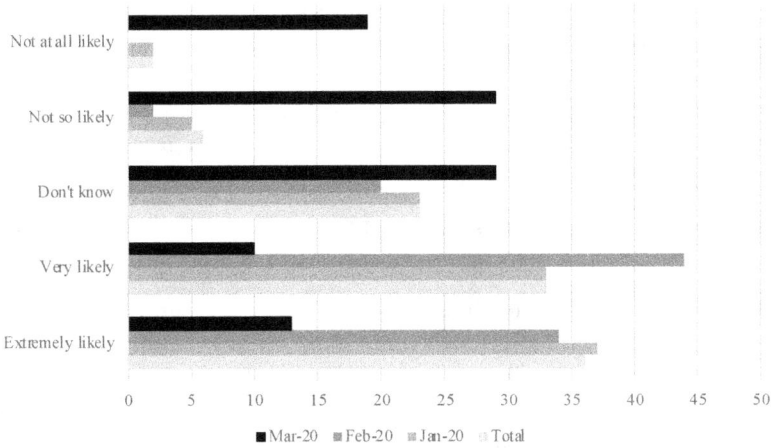

Source: Australia 2030 research Rocky Scopelliti

The imperative for organisational digital transformation is now in full swing across developed markets. According to BCG analysis[144], 52 per cent of large public companies in Europe and North America announced transformations in 2016 – a 42 per cent increase over the past 10 years. BCG's analysis sought to determine whether transformations were creating value by comparing the growth of total shareholder return (TSR) of transforming companies, with that of their respective industry. What they discovered was that only 24 per cent of transforming companies experienced greater TSG growth than their industry average over both the short term (one year), and the long term (five years +). So why are digital transformations failing?

According to McKinsey Global Institute research145, only 8 per cent of organisational leaders believe their current business model would remain economically viable if their industry continues digitising at its current course. Its research identifies five explanations of why digital strategies are failing:

1. **Organisational alignment.** Lack of a clear, wholistic organisational definition of digital and its connection to their business
2. **Value creation.** Misunderstanding the economics of digital as it disaggregates value propositions where economic rent becomes primarily redistributed in favour of customers. Scale and network effects dominate markets where economic value drives winner-takes-all economics. First movers and fast followers develop a learning advantage to out-execute competitors

3. **Competition differently.** Understanding the new economic rules that digital will apply to the broader value chain as industries increasingly become ecosystems. The platformification of industries allows players to traverse across traditional boundaries that are increasingly becoming blurred.
4. **Incumbency.** Overestimating digital attackers and underestimating the impact of a digitised incumbent with significant market share
5. **Innovation.** Not digitising their traditional businesses and innovating with new models as discussed earlier observed with digital leaders and renovators.

What then makes digital leaders and innovators succeed? There are still many industries and organisations that are yet to fully embrace all that the 3rd Industrial Revolution era has to offer technologically. A performance gap has now emerged between those organisations that invested in digital transformations and are applying digital technologies and strategies, and those that are still competing in traditional ways.

McKinsey's research reveals that digital leaders and innovators exhibited common characteristics. These were:

1. **Business transformation.** They innovate their business models in a transformational way (opposed to some incremental adjustments to strategy and core business).
2. **Scale cutting-edge technologies.** They scale up cutting-edge technologies and techniques. Apply design thinking at scale across the organisation or within business units.
3. **Decisive and significant investment.** They invest decisively and three times as much, and for the long term relative to traditional companies.
4. **Strategic context.** They see the world as interconnected ecosystems.

Despite recognising that the emergence of new businesses and delivery models would pose the greatest threats, executives continue to rely on strategies that prioritise traditional business operations and models. When it comes to workforces, 86 per cent of executives reported that they are doing everything possible to create a better workforce for this revolution, but less than a quarter are highly confident they have the right workforce composition and skills for it.

Are we innovating the right things?

Some of the most cutting-edge innovation and ingenuity taking place does not involve breakthrough technologies, but rather the creation of new business models. Historical business models have been remarkably stable, dominated by a few key ideas and upgraded by a few major variations over time.

Then the internet arrived. Business model reinvention entered a period of radical transformation. In less than two decades, we've seen network effects birth new platforms in record time. Industries and players have come and gone. We are now witnessing seven emerging business models that have the capacity to redefine business over the coming decade. So, while countless businesses are anchored by a mentality of maintaining their current business models, entrepreneurship, innovation and ingenuity is at work. Over the many years I've consulted or advised boards and leadership teams on this point, the test I've presented to them was 'is your transformation program focused on making the horse run faster......or reimaging transport? Making the horse run faster will achieve a linear improvement, however, disruption occurs when the exponential curve crosses the linear curve. As McKinsey found in its research, digital leaders and innovators innovate their business models in a transformational way (opposed to some incremental adjustments to strategy and core business). Below are just some of the business model transformations identified by the Singularity University that will create value in a new accelerated exponential way146.

The crowd economy: Crowdsourcing, crowdfunding, leveraged assets, and staff-on-demand—essentially, all the developments that leverage the billions of people already online and the billions coming online.

The free/data economy: This is the platform version of the 'bait and hook' model, essentially baiting the customer with free access to a cool service and then making money off the data gathered about that customer.

Closed-loop economies: Also referred to as the circular economy, these models will grow increasingly prevalent with the rise of environmentally conscious consumers and the cost benefits of closed-loop systems.

Decentralised autonomous organisations (DAOs): A radically new kind of company is envisaged through the convergence of AI and blockchain technologies. These autonomous systems of production are based on a set of pre-programmed rules that determine how the company operates, and computers do the rest.

Transformation economy: The next iteration of the experience economy is the idea of not only paying for an experience but paying to have your life transformed by this experience.

Need new thinking, not yesterday's logic

To use yesterday's logic to prepare leaders, organisations or policies will not set them on the right trajectory for the 4th Industrial Revolution. We need new thinking, as history has no precedent for what's to come.

In the past 10 years, price/performance has seen the costs of processing power, bandwidth and cloud infrastructure decline exponentially. This is now powering a new generation of emerging technologies that are fusing the digital, physical, environmental and biological worlds.

As Klaus Schwab, founder and executive chairman of the World Economic Forum, describes "we are at the beginning of a revolution that is fundamentally changing the way we live, work and relate to one another". He also proposes that "businesses, industries and corporations will face continuous Darwinian pressure and as such, the philosophy of "always in beta" always evolving will become more prevalent"[147].

Many leaders have formed a belief or perception that digital transformation is an on-time event that will provide some guarantee of survival. That belief has perhaps distorted expectations of performance. However, successful leadership in the 4th Industrial Revolution requires a new principle – one based on constant adaption. It's time to get your juvenescence on.

Australian professionals' lack of confidence in their organisation's survival has been rocked to its core by the impact of COVID-19. No business has avoided its impact. The age, size or industry did not make any one organisation invincible. Small businesses and restaurants closed their doors. Retailers stood down instore staff and rehired online and ecommerce staff. Many organisations struggled to adapt to the crisis that hit them. But out of this crisis, entrepreneurship suddenly emerged like restaurants providing take away and gymnasiums going online. How is this explained? It's simply 'juvenescence' at its best–the ability for leaders and their workforces to adapt to an ever-changing world around them. Equipped with a capacity to reskill, this could be Australia's secret untapped national competitive advantage.

KEY POINTS

- How we think about what we once believed 'normal', its 'rules' and what we're influenced by is all changing. This is being driven by demand for greater access to information anywhere, anytime and on any device. The power is now in the hands of every digitally connected citizen, personalised and democratised. Artificial intelligence (AI) and machine learning (ML) will become some of the biggest catalysts of change to the human workforce. Jobs are changing in the 4th Industrial Revolution, that's irrefutable.

- We need roles people can step into that will fuel our economy and launch Australia into a far more prosperous and sustainable future. Jobs like include cyber specialists, engineers, data analysts, programmers, robotic repairs, AI, ML and Internet of Things integrators and scientists. Our ability to handle unpredictable situations that require out-of-the-box thinking – critical thinking, problem solving, empathy, understanding, creativity and collaboration – are all skills that must be encouraged and developed as they are transferable and lifelong.

- Today, leaders are aware of the need to upskill, reskill and recruit employees with the required digital capabilities for this new way of work. This optimism is found whereby Australian professionals across all demographic profiles, believe that technology will create more jobs than it destroys over the 10 years.

- However, an ingenuity gap has emerged in Australia as a leadership issue, not a technological issue. Australian CEOs and their boards have higher expectations of their organisation's capability, culture and operations capacity for change, compared to management that report them. That leadership gap was tested through the COVID-19 crisis that saw organisations struggling to adapt in an agile manner, their technology and operations, workforces, supply chains and in some cases, their business models, around the restrictions being applied in Australia and around the world.

- Australian enterprises must overcome their aversion to innovation failure to encourage further digital innovation, flatten hierarchical structures, decentralise decision-making and encourage new ways of working in a collaborative manner. This can be achieved by adopting agile innovation methodologies to optimise processes through experimentation.

- Australia is reported to have one of the worst levels of work-life balance among other OECD countries. Relative to many other nations, Australia ranks in the bottom third of OECD countries when it comes to working long hours and time devoted to leisure and personal care. This became evident in the Australia 2030 research by how easily its importance declined with more than half of those respondents in March 2020 reprioritising 'reskilling and support programs' and 'not enough jobs' as their major concerns about their job prospects over the coming decade.

- There are many opportunities for Australian leaders to revisit both work-life balance and flexibility post COVID-19 when both those concerns were tested. For many who had to juggle home schooling, multiple working parents in the one household and isolation, this was quite challenging. However, what we have learned is that for many, both these can coexist and perhaps should not be traded off against salary and financial compensation in ordinary times.

- Without effective strategies developed and planned collaboratively by enterprise, academia and government, Australia risks widening its ingenuity gap. The situation means we must evolve how we develop and utilise human capital.

- We need new thinking. To many leaders, digital transformation has suggested that disruption is a one-time event, and that belief has perhaps distorted expectations of performance or survival. However, successful leadership in the 4th Industrial Revolution requires a new principle – one based on constant adaption. I refer to it as 'Juvenescence', defined as the constant state of youthfulness. It's time to get our juvenescence on. Equipping our workforces with a capacity to reskill could be Australia's secret untapped national competitive advantage.

TIPPING POINTS

1. **Closing Australia's ingenuity gap will be through a knowledge-based model, not learning based.** Australian professionals are optimistic that technology will create more jobs than it destroys over the next 10 years, providing leaders with the opportunity to rapidly respond, recover and thrive post COVID-19 and adapt to reskill workforces for the 4th Industry Revolution. However, leaders must overcome their aversion of failure through experimentation and innovation and embrace it as an integral part of the learning process.

2. **By the 2030s, universal basic income will be widespread in the developed world:** While not a technological advancement, but rather one of political will, a universal basic income will be widespread by the 2030s. Universal basic income experiments, in which people are given a regular income just to live, no strings attached, are emerging

across Europe, North America and Africa. The onset will be increased through automation and the use of artificial intelligence. For Australia, a universal basic income could profoundly change the future of work, improve the productivity of its human capital, and achieve better balances in work life.

The workforce post COVID-19

More than four out of five workers in the global workforce were estimated to have been impacted by COVID-19 through lockdown and stay-at-home measures[148]. This is the time to consider the future of work. This word cloud represents the current thinking on skills and workforces in a post COVID-19 world.

Source: Australia 2030 research Rocky Scopelliti

I'll close this chapter by sharing some of the qualitative quotes from 271 respondents who participated in the Australia 2030 research on the topics of skills, job creation, innovation and organisational survival.

IN AUSTRALIAN PROFESSIONALS' OWN WORDS

- *"Able to adapt and change quickly to changing environments"*
- *"Will be around, but will look very different"*
- *"I believe we have the right structure in place to be agile enough to move to the requirements of our customers, to value our team members and strong enough financially to be able to."*
- *"Our company is 36 years old, and it is a technology business, and has shown it can stay at the forefront of technology over a long time and with a lot of changes so I see no reason this will not continue".*
- *"Shrunk, but not forgotten".*
- *"Designed for the gig economy".*
- *"While it is one of the largest companies in Australia offering critical services, its risk is in not meaningfully pursuing a growth strategy".*
- *"Not entirely certain.. things are changing so rapidly it's hard to predict in the next 3 years – I am confident though".*
- *"We are a company that has been around for 20 years now and we are all about sustainable business and how to transform your business too. So, if we can't make it happen there is something very wrong".*
- *"Despite the current challenges, the company is well positioned to support the changing face of the social and technological landscape over the next decade".*
- *"Global and diversified, not reliant on one market".*
- *"[Company name] will be around your years to come – how big we will be or what we will look like is yet to be decided.........".*
- *"We are trying to change and transform I just don't know if we have the will for true transformation".*
- *"The new generation will stay away from ownership model and our business is ownership model".*
- *"Currently employed by a large multinational, with over 140 years of existence. Even if a downward trend emerges over the next few years, the organisation has anchor industries that they are recognised experts and I believe can be*

sustained for more than another decade".

- *"It's one of the Big Four banks. Only way I see it go down is if we get another Great Depression circa 1920s".*

- *"Impacts of global disruption are being felt and impacts in current business over the next decade could be significant unless the company transforms to deliver relevant products and services".*

- *"The size of the company, its age and global presence, doesn't mean it's going to be around tomorrow – look at Kodak...".*

- *"I have said "no so" because I believe it is unlikely the current structure and business model will survive but currently have little confidence in its ability to reinvent itself".*

- *"My company will still be around, but I suspect in a different form. i.e., it will transform its value proposition to meet the needs of a new and constantly evolving business landscape".*

- *"Our company is the 2nd largest in the industry and had proven to be adaptable to market changes. As long as they can continue to adapt, we'll be around".*

- *"COVID-19 just tipped every company on their heads – no company or person is invincible".*

- *"Our industry yes. Our ability to compete as an independent against global companies – remains to be seen".*

- *"Rate of change is increasing and hard to predict stability in current form of entities".*

- *"Yesterday we thought it would be technological development that would disrupt our business the most....then came the virus.....who predicted that?".*

CHAPTER 8

Science, Technology, Engineering, Mathematics (STEM)

"Crikey!" (Steve Irwin)

"Indigenous Australia has some of the world's earliest scientists and inventors, who have witnessed major astronomical and catastrophic events like tsunamis, meteorites, floods, and entire ice ages, and fortunately have survived to tell the story. Long before the Greeks were studying the stars Indigenous Australians were developing highly sophisticated sciences..."[149]

Luke Briscoe, Aboriginal scientist

This word cloud reflects how we feel about innovation, industry and planning for our future.

Imperative Education Industry Ambition NISA Skills Technology CSIRO Plan Science Development STEM Research REVOLUTION innovation Science

Source: Australia 2030 research Rocky Scopelliti

In this chapter we will explore two very important questions that the Australia 2030 research revealed.

1. Why don't Australian professionals have confidence that Australia's federal and state governments have effective plans in place with industries and the private sector for economic, technological, social and cultural transformation over the coming 10 years?
2. Why are Australians professionals concerned that Australia is not investing enough in technological, scientific and supporting skills development compared to other countries for the coming 10 years?

The story of Australia's technological and scientific developments acknowledges the traditional owners of the land and the contribution made by them as the world's oldest continuous living culture. Aboriginal and Torres Strait Islander people have used scientific and technological methods of data collection, such as observation and experimentation, for tens of thousands of years – long before western science and technology came to Australia. It is also important to acknowledge that from Aboriginal and Torres Strait Islander people's perspectives, there is an intricate and inextricable interconnection between the physical, chemical and biological sciences and the social sciences more widely – long before industrial revolutions of the modern world understood this. This is particularly so because of the deep and timeless relationship between Country and Aboriginal and Torres Strait Islander identities, languages, cultures and spiritualties.

For centuries, Aboriginal and Torres Strait Islander sciences have incorporated sophisticated knowledges and practices pertaining to: seasons and meteorology; astrology and astronomy; bush food, medicine and healing; natural resource management; and the physics and chemistries behind the design, production or use of tools, instruments and inventions. Some of these inventions and innovations include:

- The *Boomerang* – an invention used to hunt animals and return to sender
- The Woomera, that uses leverage to allow a spear to be thrown up to three times further
- *Thermoplastic resins* that are strong enough to bond rock to wood
- *Weirs and fish traps* that demonstrated a sophisticated understanding of engineering, physics and aquaculture in their design and use
- *Firestick farming using* precise controlled burning
- *Water bags* made from wallaby skins, which used capillary action and evaporative cooling to keep food from spoiling

- *Stone and natural glass tools* shaped into chisels, saws, knives, axes and spearheads
- *Bushfoods and medicine* used for hunting and treatments for inflammation, antiseptics and to cure infections.

Industrial revolutions

Throughout every industrial revolution of the modern world, Australia has made some of the most remarkable inventions, innovations and breakthroughs that changed the world such as:

1st Industrial Revolution (1784–1870)

- 1843 – *Grain stripper* – John W Bull invented the world's first mechanised grain stripper that was first manufactured in South Australia.
- 1856 – *Refrigerator* – James Harrison produced the world's first practical ice making machine that used the principle of vapour compression.
- 1859 – *Photolithography* – John W Osborne pioneered the art of photolithography, transferring master copies of land records and maps to 150mm square glass slides.

2nd Industrial Revolution (1870–1969)

- 1874 – *Underwater torpedo* – Invented by Louis Brennan, the torpedo could be steered by an onshore operator.
- 1889 – *The electric drill* – Arthur J Arnot patented the world's first electronic drill in Melbourne.
- 1902 – *The notepad* – JA Birchall, a Launceston stationer, decided that it would be a good idea to back loose sheets of paper with cardboard and glue them together.
- 1906 – *The feature film* – The story of the Kelly Gang was the world's first feature-length film at just over an hour in length.
- 1912 – *The Mark 1 Tank* – Lance de Mole from South Australia submitted a proposal to the British War Office for a chain rail vehicle.
- 1928 – *Electronic pacemaker* – Developed by Edgar H Booth and Mark C Lidwell, the apparatus plugged in to a power point.
- 1940 – *Zinc cream* – Faulding Pharmaceuticals developed a sun block made from zinc oxide.
- 1945 – *Penicillin-based antibiotics* – Australian scientist Howard W Florey was awarded a Nobel Prize for his role, alongside Ernest Chain and Alexander Fleming, in the development of penicillin.
- 1945 – *Hills hoist* – the rotary clothesline with a winding mechanism was

developed by Lance Hill. The original was designed in 1926 by Gilbert Toyne in Adelaide.

- 1953 – *Solar hot water* – Developed by the CSIRO led by Roger N Morse.
- 1957 – *Stainless steel braces* – Developed by Percy R Begg and Arthur Wilcock, the system allowed for gradual adjustments rather than the forced methods previously used to straighten teeth.
- 1958 – *Black box flight recorder* – Dr David Warren developed the black box voice and instrument recorder.
- 1961 – *Ultrasound* – David Robinson and George Kossoff from the Australian Department of Health developed the first practical water path ultrasonic scanner.
- 1965 – *The wine cask* – South Australian Thomas Angove invented the cardboard box housing a plastic container.

3rd Industrial Revolution (1969–today).

- 1972 – *Orbital engine* – Ralph Sarich of Western Australia invented the system that uses a single piston to directly inject fuel in to orbiting chambers.
- 1979 – *Bionic ear* – Professor Graeme Clark of the University of Melbourne invented what we know as the cochlear implant.
- 1983 – *Winged keel* – Ben Lexon designed the keel that memorably helped Australia to win the America's Cup that had been held by the New York Yacht Club for 132 years.
- 1984 – *Frozen embryo baby* – the world's first frozen embryo baby was born in Melbourne.
- 1992 – *Wifi* – Developed and patented by CSIRO researchers who discovered radio waves that echo off indoor surfaces.
- 1995 – *Gene slicing* – A CSIRO team led by Professor Peter Waterhouse discovered that double-stranded RNA was the trigger for gene editing.
- 2006 – *Cervical cancer vaccine* – Professor Ian Frazer from the University of Queensland created a preventative vaccine for cervical cancer.
- 2012 – *Quantum bit* – A team of Australian scientists, led by Andrew Dzurak of the University of Sydney and Andrea Morello of the University of NSW, built the first quantum bit using a single phosphorus atom implanted into a silicon chip.
- 2018 – *Modular self-fit hearing aid* – Led by Professor Peter Blamey and Dr Elaine Saunders, in collaboration with the Victorian government, RMIT and Swinburne University of Technology, the team released the first hearing aid that allows users with severe dexterity issues to self-manage their hearing aids.

At the cutting edge of scientific and technological development

It's important to acknowledge the contribution this nation has had, and continues to have, through the Commonwealth Scientific and Industrial Research Organisation (CSIRO). As referred to in chapter 6, economist Mazzucato's research clearly highlights the importance and significance of the role of the government in innovation. The CSIRO began as the Advisory Council of Science and Industry in 1916 and was subsequently born in 1949, post-Second World War under the *Science and Industry Act of 1949*. That Act defined CSIRO's purpose, which was to carry out scientific research for assisting Australian industry; furthering the interests of the Australian community; contributing to the achievement of Australian national objectives or the performance of national and international responsibilities of the Commonwealth; and, any other purpose determined by the Minister. Importantly, to encourage or facilitate the application or utilisation of the results.

The CSIRO has a legacy that includes the invention of fast Wi-Fi, Aerogard, extended wear contact lenses, BARLEYmax™, self-twisting yarn and polymer banknotes. Things we take for granted, particularly wifi during the COVID-19 lockdown. Today the CSIRO is trying to find the first gravitational waves in space, growing gluten-free grains, 3D-printing body parts and pioneering new renewable energy sources and, through its laboratories in Victoria, seeking to develop a vaccine for COVID-19, just to name a few. The future of Australian science and technology is very bright. Australia has a very long history of original, ingenuous and cutting-edge thinkers.

Why begin the chapter with a historical reflection on Australia's capacity for innovation? Simply because as I write, both the public and private sectors researchers from CSIRO, the University of Queensland, CSL, Mesoblast and Vaxine are working around the clock to develop COVID-19 vaccines and treatments in record time to save the lives of the critically ill and put an end to the pandemic not just in Australia, but indeed around the world. The CSIRO has commenced the first stage of testing potential vaccines for COVID-19 through its high containment biosecurity facility at the Australian Centre for Disease Preparedness, formerly known as the Australian Animal Health Laboratory. It's important to acknowledge that the CSIRO was the first research organisation outside of China to generate sufficient stock of the virus. In February 2020, the CSIRO was the first organisation in the world to confirm that animals react to SARS-Cov-2—the virus that causes COVID-19.

In chapter 1, the Australia 2030 research revealed that Australian professionals trust the government the least when it comes to controlling their best technological and scientific interest over the coming decade, but in chapter 6, we discussed how that trust was primarily placed in the hands of 'academic and research institutions' and 'large public technology or scientific institutions. In chapter 6, we noted that the Australia 2030 research found that Australians are concerned that the country is not investing adequately in technological, scientific and skills development compared to other countries over the coming years. The advent of COVID-19 over the study period (January–March 2020) exacerbated that sentiment as concerns increased from 55 per cent in January to 95 per cent in March. During March, as the death toll increased, the lockdown of the entire population, enforcement of social distancing measures, isolation and other measures impacted all Australians (see Exhibit 8.1).

Exhibit 8.1: Q. Are you concerned (or not) that Australia is (or isn't) investing enough in technological, scientific and skills development compared to other countries for the coming 10 years? (Select 1 Response)

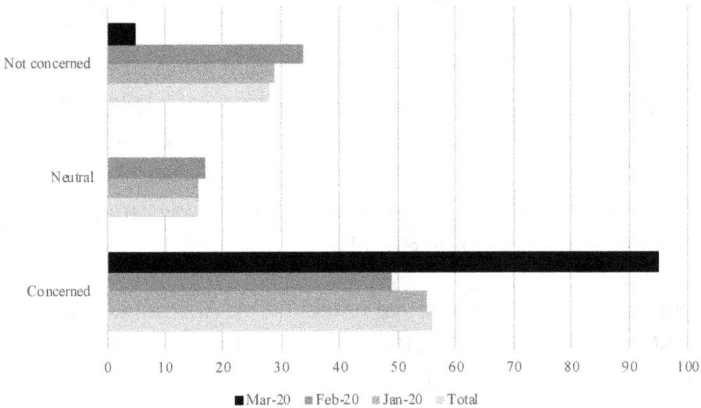

Source: Australia 2030 research Rocky Scopelliti

Does this reflect a much greater confidence or optimism deeply seeded in our pioneering history that is underutilised by governments policies, or the now redundant leadership model that during COVID-19 was replaced through the systems leadership model, but remains to be instilled? Does this also reflect attitudes that have seen policy settings in the last decade interlocked into the electoral cycle, as we saw in chapter 1, at the expense of our future? Let's take a closer look.

Addressing Australia's innovation inefficiency

The overarching need for innovation, investment, research and development is clear. Without them Australia will not experience improvements in our standard of living, national productivity, education, health and much more. As seen through Australia's history, innovation, ingenuity and inventiveness is a significant part of the fabric of who we are today. However, in our modern history Australia has failed to realise the full potential of many of those innovations that were commercialised overseas. A classic and well-cited example is the black box flight recorder. As discussed in the last chapter, innovation will be key to driving future productivity growth, as well as developing new industries, new companies centred on new and emerging technological and scientific skills. However, Australia's innovation capacity is constrained by a risk averse leadership culture towards experimentation and innovation failure.

According to the Global Innovation Index 2019 report, Australia's innovation score places it 22nd in the world – a drop of two positions from the prior year's ranking[150]. Regionally, that places Australia behind Singapore, South Korea, Hong Kong, China and Japan. The major deficiencies in Australia's innovation rankings were reported to be in the areas of: overseas research and development financing (84th), per-worker GDP growth (77th); percentage of graduates in science and engineering (76th); government funding per student (69th); GDP per unit of energy use (67th); exports of national feature films (56th); culture and creative services (56th); overall investment climate (51stt) and knowledge absorption (50th).

Australia's efficiency in translating innovation inputs into outputs is comparatively lower than many other countries. Indeed, Australia's Innovation Efficiency Ratio (innovation input/innovation output) ranks only 76th in the world. This is negatively influenced by much lower ranking in innovation outputs (31st) than in innovation inputs (11th).

However, there are examples of where collaboration between government and industry produces improved innovation output (see Case Study Agricultural Innovation). What this highlight is that when Australia moved to remove most of its agricultural trade barriers, the quid pro quo was a system of RDCs and levies to fund agricultural research and develop. Grower levies are collected, the government matches the levy fund on a 1-for-1 basis. The RDCs use that to fund agricultural research and development for their community. The end result is the single largest portfolio of research and development in the world.

Australia spends ~$3 billion on Rural research and development per annum – about $1.6B of that is government funded151.

The metrics show it produces a massive body of high-quality targeted research – possibly the most significant in the world. But the vast majority either stays stranded by intellectual property or goes overseas for commercialisation before (eventually) having a commercial impact in Australia. As a result, we have a low efficiency and a very slow diffusion rate despite the massive spend.

Case Study: Agricultural Innovation

Following a collaborative stakeholder engagement process, Ernst and Young has developed a shared vision for agricultural innovation. Realising the vision for Australia's agricultural innovation system will allow us to harness the power of knowledge to make our food and fibre systems more competitive, prosperous and sustainable.

The vision report is titled *Agricultural Innovation—a national approach to grow Australia's future* and was launched on 5 March 2019. The report makes recommendations aimed at benefiting all participants in Australia's agricultural innovation system, including researchers, research and development corporations (RDCs), industry representatives, producers, processors, investors, government agencies and companies across the start-up, accelerator and incubator communities.

There are five recommendations in the report:

- strengthening leadership for strategic direction, but also for improving connections, collaboration, and culture
- balancing funding and investment to solve short-term challenges as well as targeting transformational and cross-commodity outcomes
- establishing world-class innovation practices including disruptive thinking, ambition and entrepreneurship to maximise opportunities from our investments
- strengthening the regions to maximise innovation uptake and provide regions with a greater role in national priority-setting
- establishing the next generation innovation platform for our data, physical infrastructure, and regulatory environment.

A world class agricultural innovation system will help Australia to reach the National Farmers Federation's target for a $100 billion sector by 2030.

Source: Department of Agriculture, Water and the Environment

A plan to address Australia's innovation performance

To address Australia's deteriorating innovation performance, in December 2015, the Turnbull government established the National Innovation and Science Agenda (NISA). NISA's independent statutory board of 15 entrepreneurs, investors, researchers and educators was tasked to produce a strategic plan to advise policy makers on how to accelerate innovation and optimise Australia's innovation system through to 2030. In its report titled *Australia 2030: Prosperity through Innovation* (the 2030 Plan)[152], NISA made 30 recommendations that underpin five interrelated strategic policy initiatives (see Exhibit 8.2).

Exhibit 8.2: Five imperatives for the Australian innovation, science and research system

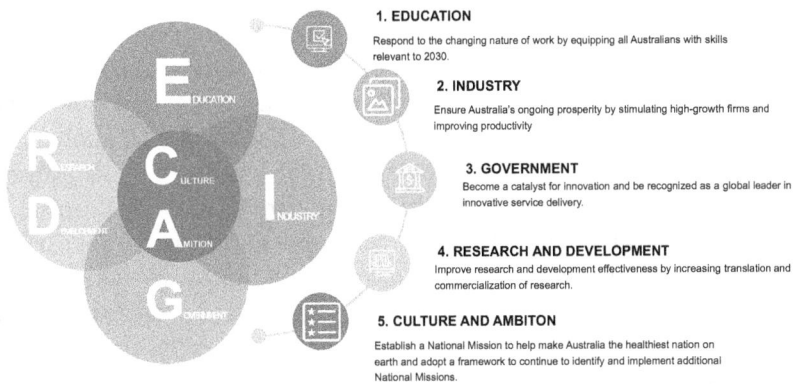

1. EDUCATION
Respond to the changing nature of work by equipping all Australians with skills relevant to 2030.

2. INDUSTRY
Ensure Australia's ongoing prosperity by stimulating high-growth firms and improving productivity

3. GOVERNMENT
Become a catalyst for innovation and be recognized as a global leader in innovative service delivery.

4. RESEARCH AND DEVELOPMENT
Improve research and development effectiveness by increasing translation and commercialization of research.

5. CULTURE AND AMBITON
Establish a National Mission to help make Australia the healthiest nation on earth and adopt a framework to continue to identify and implement additional National Missions.

Source: NISA

Let's take a closer look at each of these.

Imperative 1 - Education

'ISA's vision is that Australia has a world-leading education system that equips all Australians with the skills and knowledge relevant to 2030. Realising this vision is the first imperative of this plan because providing a world-class education is fundamental to Australia being an innovative and fair country. Education determines the capability of workers and entrepreneurs, and therefore the economy's productivity and innovation capacity. Education also shapes Australians' life opportunities.'

Education as the first imperative is significant as it also acknowledges that for more than 20 years, business, government, media and our institutions have drawn our attention to the science, technology, engineering and mathematics (STEM) problems facing Australia. As also highlighted in the last chapter, our capacity to reskill our workforces is a major contributor to Australia's ingenuity gap.

Despite total government funding increasing over the past decade from approximately $13,400 per student in 2010 to $15,200 per student in 2016, we saw funding per student decline significantly for science by $1,000 and mathematics by $1,300. So according to the RBA's inflation calculator, that's an increase of $5.72 per student and overall the government generously increased student funding by 0.04 per cent over six years. The challenge Australia has is that its school system performance has declined in the past decade, both relative to other countries and in real terms.

The Department of Education, Skills and Employment identified 108 STEM occupations based on the ABS occupation classification. This list has been established through an analysis of the skill and task composition of occupations to determine whether there are scientific, technological, engineering or mathematical requirements for job seekers considering their career pathway. While STEM is related primarily to the field of education, it is also an important lens to analyse skill requirements for occupations in the Australian labour market. As there is no commonly agreed categorisation of STEM occupations, the Department of Education, Skills and Employment has reviewed other research and information sources in the establishment of the STEM occupation list[153].

To address these issues, the NISA report made the following recommendations and actions:

1. Strengthen training for pre-service and in-service teachers
2. Better prepare students for post-school science, technology and numeracy
3. Raise student ambition and achievement in literacy and numeracy
4. Review the vocational education and training system
5. Continue and strengthen reforms to the vocational education and training system.

While addressing the school's system issues relating to STEM for new workforces, the more immediate priority is the reskilling of Australia's existing

13 million strong workforce[154]. As highlighted in the last chapter, the Australia 2030 research shows that 39 per cent of Australian professionals are concerned about 'reskilling and support, 'not enough jobs' and 'lack of technological skills' when it comes to their job prospects over the coming decade.

The wonderful 2016 movie *Hidden Figures* reminds us of just how far we have come in a relatively short time, but yet, from a gender diversity perspective, how far we've got to go. Sadly, Katherine Johnson passed away in February this year. She was an African American mathematician known for accuracy in computerised celestial navigation. She made significant contributions to the USA's aeronautics and space programs with the early application of computers at NASA. Katherine began work in 1953 with the National Advisory Committee for Aeronautics (NACA) (superseded by NASA in 1958), working in a pool of women performing mathematical calculations. She referred to the women in the pool as 'virtual computers who wore skirts.' Today, that time-consuming work can be done in milliseconds with an in-memory computer.

Just as in-memory computing transformed calculating by hand (and made jobs like Katherine's much easier), digital technologies are transforming the way we work today–and making our day-to-day activities more efficient.

History does indeed demonstrate that labour markets adjust to changes in demand for labour from disruptive technological advancements associated with industrial revolutions. So, we need to rebalance our thinking towards questions such as *what new jobs will be required to support the next industrial revolution and how do we reskill and create new skills (and jobs) to support the future demand of labour?* A majority of Australian professionals (52 per cent) believe that technology will definitely create more jobs than it destroys over the next 10 years. Unsurprisingly, Millennials felt this more than other demographic groups.

Interestingly, that belief skyrocketed in March (90 per cent) as COVID-19 unfolded, perhaps heightened by our reliance on technology for remote working, social contact and its role in fighting the disease. But it's not just the workforce itself that we need to consider, but the very real possibility that through this event, we may need to redefine the nature of work and its relationship to reward (See Exhibit 8.3). It makes you wonder how over a decade of anti-science rhetoric from various political leaders has impacted the baseline figure for trust in science? Imagine how much higher it would have been. Events where science plays a key role (such as dealing with COVID) rapidly peel away that ideologically driven sentiment.

Exhibit 8.3: Q. In your opinion, do you believe technology will create more jobs than it destroys over the coming 10 years? (%)

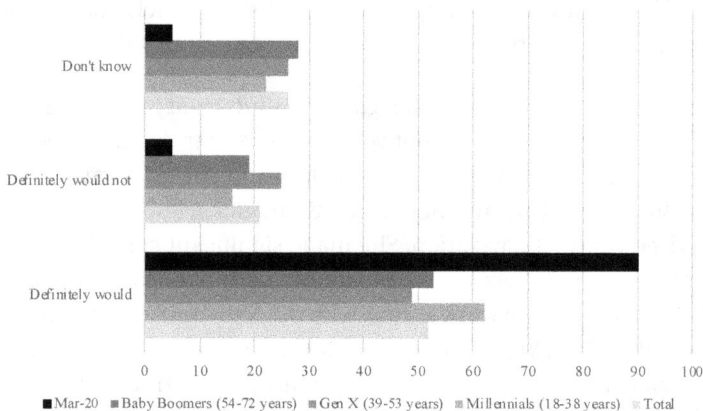

■ Mar-20 ■ Baby Boomers (54-72 years) ■ Gen X (39-53 years) ■ Millennials (18-38 years) Total

Source: Australia 2030 research Rocky Scopelliti

Imperative 2 – Industry

'ISA's vision is that by 2030 Australia will accelerate growth and exports of Australian businesses by strengthening a competitive and productive business environment.'

The shift towards a more service-based economy

Over the decade, we have seen a progressive shift in Australia's distribution of GDP across economic sectors. In 2018, agriculture contributed around 2.46 per cent to the GDP of Australia, 24.12 per cent came from industry, and 66.56 per cent from the services sector. However, Australia closed this decade with an economy that has posted its equal slowest annual growth since the year 2000. The economy grew by 1.4 per cent over 2019, the worst annual growth recorded in the aftermath of the global financial crisis in September 2019.

Over the 30 years between May 1989 and May 2019, employment increased by 5.2 million (or 66.9 per cent)[155]. Over that same period employment grew in 16 of the 19 broad industries, with the majority of growth recorded across service industries (see Exhibit 8.4).

Exhibit 8.4: Change in industry share of employment, 30 years to May 2019, by broad industry ('000)

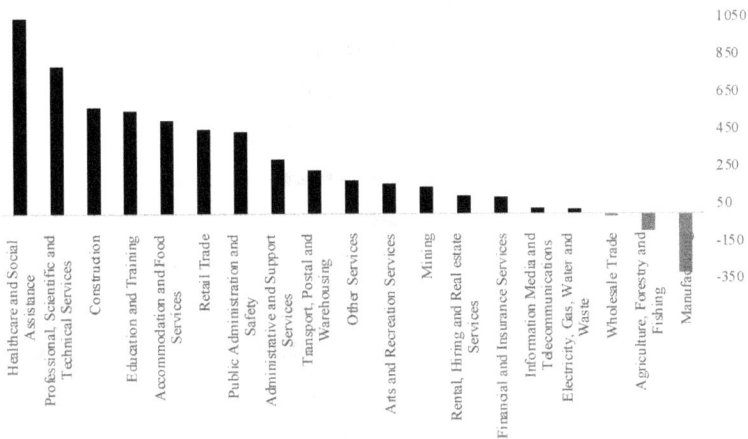

Data Source: ABS, Labour Force, Australia, Detailed, Quarterly, cat. no. 6291.0.55.003, trend.

The largest increase in employment was recorded in the health care and social assistance industry (up by 1,047,300 or 163.2 per cent), with growth in the industry underpinned in the longer term by ongoing population growth and an ageing population. The strong growth recorded over the past 30 years saw the industry move from the third largest employing industry to the top employing industry. The industry's share of total employment also rose from 8.4 per cent to 13.1 per cent.

Employment is projected to rise in 17 of the 19 industries over the five years to May 2023[156]. Around two thirds of new jobs during this period are expected to come from four industries (see Exhibit 8.5).

1. Health care and social assistance (up by 250,300, or 14.9%).
2. Construction (118,800, or 10.0%).
3. Education and training (113,000, or 11.2%).
4. Professional, scientific and technical services (106,600, or 10.2%).

Exhibit 8.5: Projected employment growth, industry share (%) of total new jobs

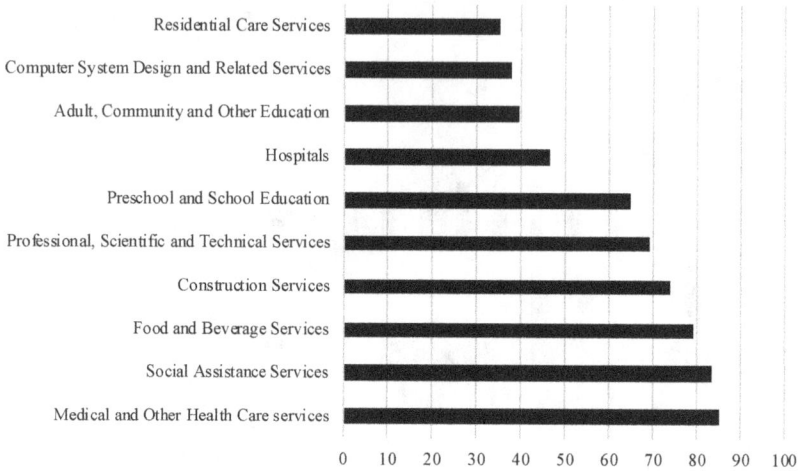

Data Source: Australian Government Labour Market

In response to those shifts in the economy and labour market, the federal government will invest approximately $2.4 billion[157] in growing Australia's research, science and technology capabilities over the next 12 years to support a stronger and smarter economy.

This funding will include:

- Supercomputers
- World-class satellite imagery
- Improving GPS across Australia
- Upgrading the Bureau of Meteorology's technology platform
- Establishing a national space agency
- Leading research in artificial intelligence.

Australia's business expenditure on research & development (BERD) is low compared with OECD countries. Its BERD is under 1 per cent of GDP, less than two-thirds of the OECD average of 1.67 per cent as a share of GDP but Australia invests a higher proportion of GDP in R&D than Italy, Canada, New Zealand, Portugal, Spain and Greece (see Exhibit 8.6).

Exhibit 8.6: Business expenditure on research and development, FY17−18 (% of GDP)

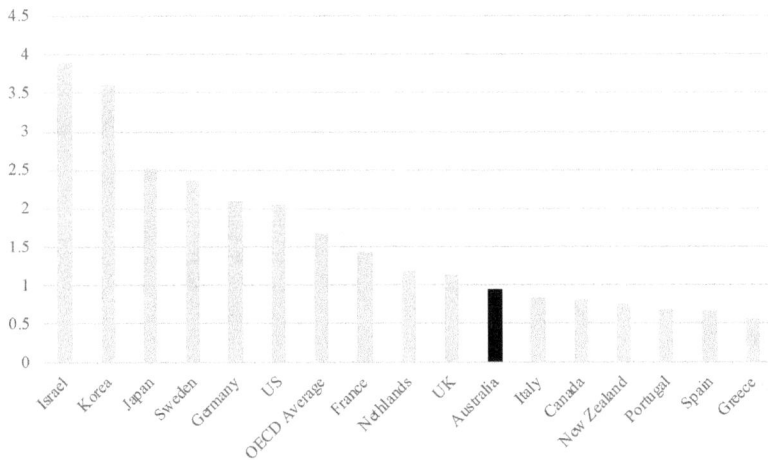

Data Sources: OECD; ABS

According to a report prepared for the Office of Innovation & Science Australia[158], at an industry level, the two sectors that contribute most to BERD in other advanced economies contribute significantly less in Australia. Manufacturing drives 70 per cent of BERD in a group of peer economies − Canada, the US, UK, Korea, Japan, and Germany − but only about a quarter of BERD in Australia. Similarly, the information media and telecommunications (IMT) sector accounts for 24 per cent of BERD in the OECD, but just 3.5 per cent in Australia. The questionable explanation provided was that in part, the manufacturing and IMT sectors are less BERD-intense in Australia than they are in the rest of OECD (see Exhibit 8.7)

Exhibit 8.7: International BERD intensity by sector, 2017–18 (% of GVA)

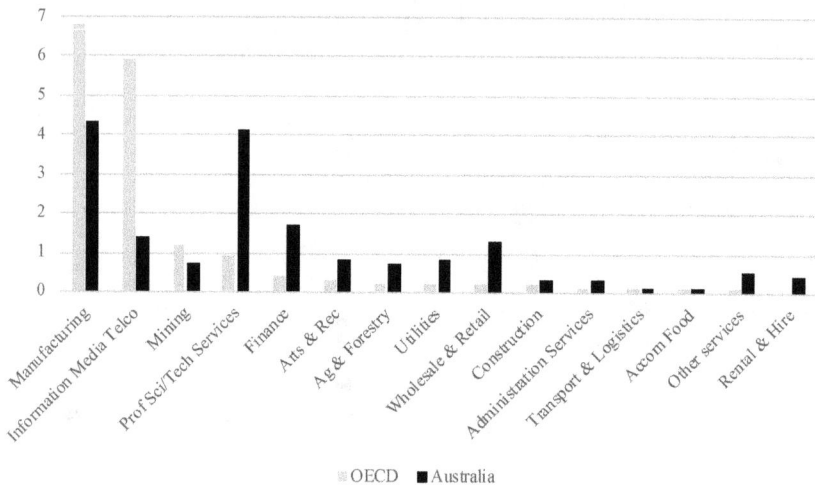

Data Sources: OECD; ABS

As shown in Exhibit 8.7, Australian manufacturers invest 4.3 per cent of Gross Value Added (GVA) in R&D; elsewhere in the OECD, manufacturers invest 6.8 per cent. Australian IMT firms invest 1.4 per cent of GVA in R&D, but on average in the OECD IMT firms invest 5.9 per cent of GVA.

According to the same report, the challenge is for many more Australian firms to invest in innovation rather than in R&D alone. Almost a third of Australian firms invest in some form of innovation, while just 5.8 per cent of firms invest in R&D. Of the firms that do invest in R&D, four in five spend more than half their total innovation budget on non-R&D activity. When Australian firms invest in non-R&D innovation, many of them invest in technology. Almost one in five firms purchased machinery or equipment for innovation in 2017. Around 14 per cent of firms spent money on internal business re-organisation for innovation. About the same proportion of firms report spending on marketing or training related to innovation (see Exhibit 8.8).

Exhibit 8.8: Share of all firms engaging in type of expenditure for innovation purposes (% of firms 2016–17)

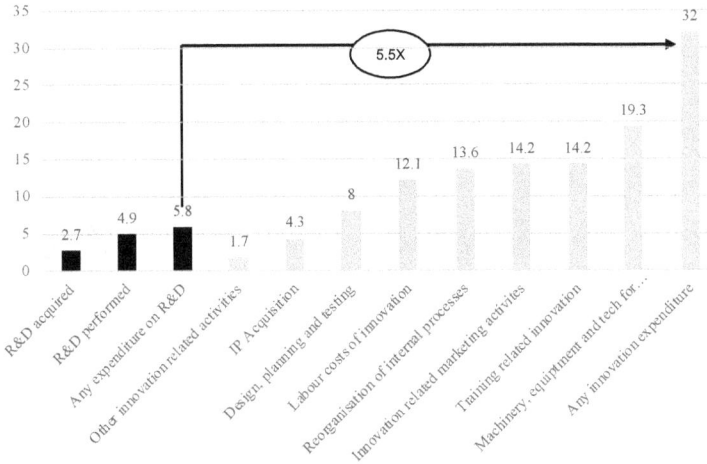

Data Source: ABS

To address the issue of Australia's business expenditure on research and development, the ISA report made the following recommendations and actions:

1. Reverse the current decline in business expenditure on R&D by improved targeting of government support
2. Enhance efforts to help young firms access export markets
3. Prioritise investment in artificial intelligence and machine learning
4. Ensure healthy competition in knowledge intensive industry sectors
5. Strengthen efforts in talent attraction and skilled migration.

Imperative 3 – The government

'ISA's vision is that by 2030, Australian Governments will facilitate innovation through the regulatory and policy environment; procurement and major programs and projects; and through role modelling innovation through service delivery.'

The government plays a major role in the Australian economy, which was highlighted in 2019 whereby public investment and government spending overall accounted for nearly 70 per cent of the growth in the economy for that year (see Exhibit 8.9).

Exhibit 8.9: Impact of Australian Government investment and spending on annual GDP growth (June Quarters %)

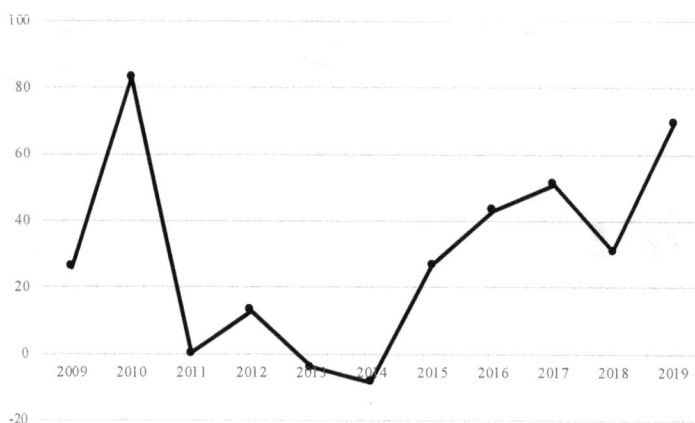

Source: ABS

Many organisations including small medium enterprises (SMEs) rely upon government contracts. The past decade has seen a significant increase in the total value of procurement contracts by the government. In 2018-19 there were 78,150 contracts published on AusTender, with a combined value of $64.5 billion most of which (94.1 per cent) contract volume was below $1 million in value (see Exhibit 8.10)[159].

Exhibit 8.10: Total value of procurement contracts by financial year ($M)

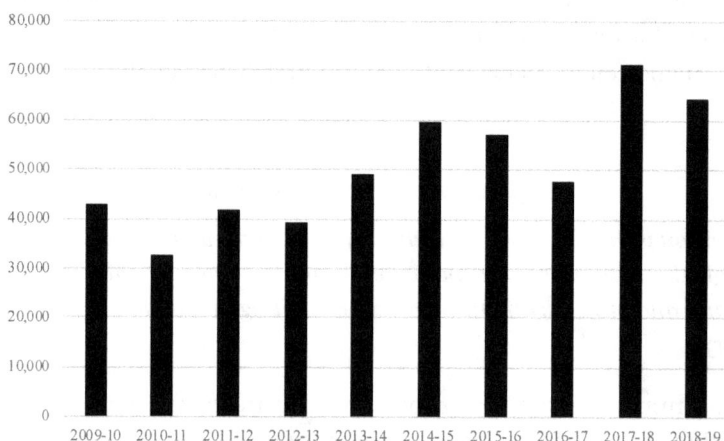

Source: Australian Government Department of Finance

Most of the volume of contracts (95.7 per cent) were awarded to businesses with an Australian address. For the 2018–19 financial year, 80.5 per cent of the value of those contracts supported the following industries:

1. Commercial and military and private vehicles and their accessories and components (25.5 per cent of total value $16,433 million)
2. Management and business professionals and administrative services (20.1 per cent of total value $12,940 million)
3. Engineering and research and technology-based services (15 per cent of total value $9,680 million)
4. Health care services (7.2 per cent of total value $4,642 million)
5. Building and construction and maintenance services ($6.6 per cent of total contract value $4,265 million)
6. Information technology broadcasting and telecommunications (6.1 per cent of total contract value $3,930 million).

At a departmental level, 80 per cent of total value spent in 2018–19 was from four departments – defence 67 per cent, human services 6 per cent, home affairs 3.8 per cent and taxation office 3.3 per cent.

In 2018, the government announced an additional commitment to source at least 35 per cent of contracts valued up to $20 million from SMEs (defined as an Australian or New Zealand company with fewer than 200 full-time equivalent employees) – up from 26 per cent in the past year. Over the 2018–19 financial year, 84 per cent of federal government suppliers were estimated to be SMEs that received contracts totalling $16.7 billion.

The public sector employs approximately 2 million people (June 2019) across a wide range of departments and agencies, which represents approximately 15 per cent of the Australian workforce. According to a UBS economist, 80 per cent of new jobs created in 2019 were in the public sector[160] (see Exhibit 8.11).

Exhibit 8.11: Employment growth – annual average (%)

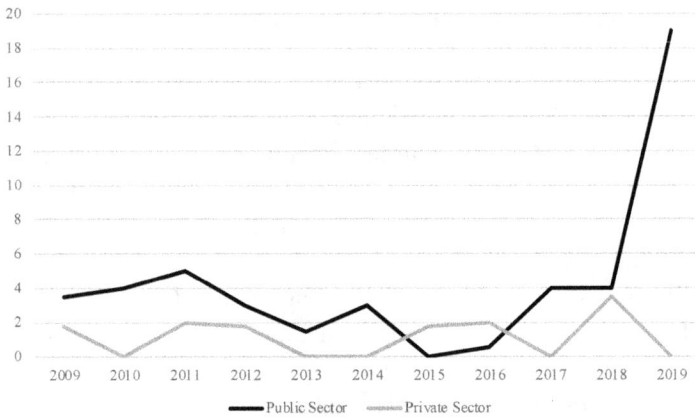

Data Source: UBS

These indicators illustrate how important government is to the economy, as without its investment, spending and employment, the economy would not have grown in 2019.

The ISA report made the following recommendations and actions:

1. Create a more flexible regulatory environment that fosters innovation
2. Encourage social innovation investment across Australia
3. Improve provision and use of open government data
4. Grow government procurement from SMEs to 33 per cent by 2022
5. Increase the use of innovative procurement strategies
6. Maximise the spill over benefits of major government programs
7. Deliver greater government savings from digitising service deliver
8. Review the public service emphasising improved capability to innovate.

Imperative 4 – Research & Development

'ISA's vision for Australia's R&D sector is to maintain the excellence that has become its hallmark, while increasing the incentives for collaboration and commercialisation. Despite significant and positive policy changes that have been made in this area in recent years, ISA believes more can be done to break down barriers between the research sector and industry, and to build stronger connections between the two'

The past decade has seen a 3 per cent decline in expenditure on R&D by Australian businesses (BERD). During 2017–18 that expenditure was $17,438 million. More positively however, was the 33 per cent increase in human resources devoted to R&D, which in 2017–18 totalled 74,991[161]. When we look at gross expenditure on R&D (GERD) over the decade across different sectors – the government, higher education and private non-profit sectors, we can clearly see that the major declines in R&D investment have come from the business and government sectors (see Exhibit 8.22)

Exhibit 8.12 Gross expenditure on R&D by sector ($ Millions)

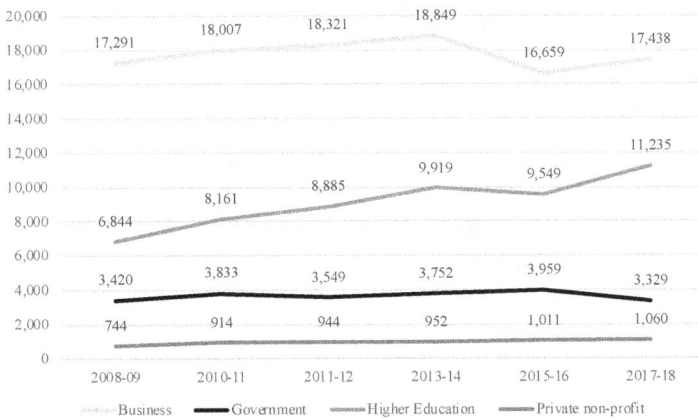

Data Source ABS

As a proportion of Australia's GDP, GERD has also decreased over the decade from 2.25 per cent in 2008–09 to 1.79 per cent in 2017–18.

So, let's come back to the question of whether Australian professionals concerns toward under investment by government and lack of future plans over the coming decade explained by the policy settings of the past decade interlocked into the electoral cycle, at the expense of our future? In the federal government's 2019–20 budget, there was little mention of research and development. Year on year comparisons, indicate a decline by $1.35 billion associated with the controversial changes to the research and development tax incentives put on hold. The explanation of this from the government was that this reduction is a one-off adjustment[162]. According to the former chairman of ISA, Bill Ferris, to get to where Australia needs to be by 2030, the government needs to double its investment in research and development and the business sector triple its investment[163].

Should Australia prioritise sectors that can lead innovation, and if so, which sectors?

As the report to the Office of Innovation & Science Australia highlighted, Australia's industry mix differs from that of other advanced economies, driving much of the overall difference in economy-wide R&D and intangibles intensity. The two key sectors that drive much of BERD globally – manufacturing and IMT – are smaller and less BERD intense in Australia. Meanwhile, the country's industry mix and innovation intensity in services – particularly business services – are above international averages. These two facts suggest that Australia may not be reaching its full potential: it may be able to improve its innovation effort by growing its manufacturing and information, media and technology sectors and, encouraging innovation in them, and by capitalising further on innovation capacity in business services innovation. The 4th Industrial Revolution provides that opportunity through emerging technologies such as advanced robotics, artificial intelligence, 5th generation mobile networks, biotechnology and material science.

A key choice for Australia is therefore whether to take a more strategic, targeted approach to accelerating innovation investment in high-potential sectors, and if so, whether the focus should be on newer sectors where Australia may have an emerging competitive advantage, or on playing "catch-up" in the most BERD-intensive sectors globally. The options advanced in that report include:

1. Focusing on business services by encouraging innovation in sectors where Australia may have a genuine opportunity to be a global innovation leader, as in professional, scientific and technical services and financial services, which are Australia's largest and most BERD-intense sectors relative to global peers.
2. Growing the manufacturing and IMT sectors, which drive a significant portion of innovation and investment globally. Australia could also choose to target segments of the industries where it already has a strong capability and value proposition, such as B2B software in IMT. These two strategies each present opportunities and risks. Focusing on growing the manufacturing and IMT sectors is not unprecedented internationally, but it would entail entering challenging global markets.
3. Focusing on business services would leverage Australia's existing innovation strengths and sector mix, but these sectors can be less easily traded and so may not offer the same international growth opportunities as other approaches. In assessing options, policymakers will need to evaluate the prospects for a strong return at the firm level and more broadly to R&D and non-R&D innovation.

To address these, the ISA report made the following recommendations and actions:

1. Introduce a collaboration premium on tax offset to incentivise collaboration
2. Evaluate scaling up industry higher degree by research placement programs
3. Evaluate the impact of recent changes to incentivise collaboration in 2022
4. Increase commercialisation capability in research organisations
5. Develop and release an Australian innovation precincts statement
6. Establish secure, long-term funding for national research infrastructure
7. Maintain a long-term policy commitment to greater gender diversity
8. ISA to monitor availability of risk capital to high-growth businesses.

Imperative 5 – Culture & Ambition

ISA's vision is that Australia seizes the opportunity to add a more ambitious chapter on innovation to our evolving national stories. We see a future as an innovation-strong nation that is also innovation proud. We believe the Australian Government has a strategic opportunity to use "National Missions" – large-scale initiatives catalysed by Governments that are designed to address audacious challenges – to accelerate Australian innovation and encourage more collaboration across the innovation system.

I began this chapter by reflecting on the traditional owners of the land, and the most incredible innovations that have shaped humanity, its survival and indeed Australia. From Aboriginal and Torres Strait Islander people's perspectives, there is an intricate and inextricable interconnection between the physical, chemical and biological sciences and the social sciences. This is particularly so because of the deep and timeless relationship between Country and Aboriginal and Torres Strait Islander identities, languages, cultures and spiritualties. And so, we should reflect on that understanding as the 4th Industrial Revolution is the augmentation between out physical, biological, environmental and now our digital worlds.

The past decade began with so much promise for technology. The Rudd government kicked off the decade with the bold, overly ambitious, and underfunded plan to roll out ultra-fast fibre broadband to every property that could receive it through the National Broadband Network (NBN). As the largest infrastructure project in Australia's history, the NBN was the subject of significant political contention and a major election issue throughout the decade. As initially proposed in 2009, wired connections would have provided up to 100 Mbit/s (later increased to 1000 Mbit/s), then decreased to a minimum of 25 Mbit/s in 2013 after the election of the Abbott government. The NBN cost jumped from $29.5 billion before the 2013 federal election, to $46–$56 billion afterwards but by late 2018, the estimated final cost was $56 billion. When it comes to the big bets or national missions, we must not make the same mistakes as evidenced with the NBN.

To address these, the ISA report made the following recommendations and actions:

1. Establish a national mission to help make Australia the healthiest nation on earth
2. Adopt a framework to continue to identify and implement additional national missions.
3.

So, while Australian professionals believe government does not have a plan – does Australia have a strategic plan for innovation to 2030. Well the answer to that is the easy part and it's a definite yes. So, if we have a plan, why do we feel we don't? Let's remind ourselves of where we started. In 2015, a new prime minister Malcolm Turnbull tried to reenergise the pioneering Australian spirit by announcing the formation of NISA. Now, this came on the back of the Abbott government ripping $3 billion out of the research and innovation system and did not even have a science minister. And while the Turnbull government's $1 billion add back at the time of announcing NISA left our nation's innovation and ingenuity with a deficit at the expense of recovering the fiscal deficit, throughout this period we also saw the sixth industry minister appointed and what residual budget was left for research and development was spread across 13 portfolios and 150 budget lines. But Australian professionals' concerns that we are not investing adequately in our technological and scientific future is rightfully founded. More progressive countries on the WEF's Global Innovation Index, bet long not short at such advance beyond the electoral cycle.

KEY POINTS

- Pioneering innovation is deeply connected to our history and culture. The story of Australia's technological and scientific developments acknowledges the traditional owners of the land and the contribution made by them as the world's oldest continuous living culture. Aboriginal and Torres Strait Islander people have used scientific and technological methods of data collection, such as observation and experimentation, for tens of thousands of years. Aboriginal and Torres Strait Islander people also know there is an intricate and inextricable interconnection between the physical, chemical and biological sciences and the social sciences— particularly because of the deep and timeless relationship between Country and Aboriginal and Torres Strait Islander identities, languages, cultures and spiritualties. As such, for centuries, they have incorporated sophisticated knowledges and practices of science pertaining to: seasons and meteorology; astrology and astronomy; bush food, medicine and healing; natural resource management; and the physics and chemistries behind the design, production or use of tools, instruments and inventions.

- The Australia 2030 research found that Australian professionals are concerned with the underinvestment in technological, scientific and skills development compared to other countries for the coming years. And rightly so. According to the Global Innovation Index 2019 report, Australia's innovation score places it 22nd in the world and regionally behind Singapore, South Korea, Hong Kong, China and Japan. Australia's efficiency in translating innovation inputs into outputs is comparatively lower than many other countries ranking it 76th in the world.

- To address Australia's deteriorating innovation performance, the Turnbull government established the National Innovation and Science Agenda (NISA). In its report titled *Australia 2030: Prosperity through Innovation*, NISA made 30 recommendations that underpin five interrelated strategic policy initiatives in the areas of education, industry, government, research & development and culture & ambition.

1. **Australia achieves top quartile Innovation Efficiency Ratio.** To achieve that, Australia will need to improve its innovation outputs. This can only be achieved through increasing its overseas research and development financing, improving per-worker GDP growth, increasing the graduation of students in science and engineering, increasing the funding per student and overall investment climate.

2. **Effective execution of the NISA 2030 Plan.** Performance reporting annually against each of the 30 recommendations.

The importance of science, technology, engineering and mathematics (STEM)

Most Australian professionals do not have confidence that Australia's governments have effective plans in place with industries and the private sector for Australia's economic, technological, scientific, social and cultural transformation over the coming decade. This word cloud represents the NISA 2030 Plan and its 30 recommendations.

Data Source: NISA

I'll close this chapter by sharing some of the qualitative quotes from 382 respondents who provided explanations for their choice.

IN AUSTRALIAN PROFESSIONALS' OWN WORDS

- *"For the past 10 years our Governments (on all levels) have repeatedly been too slow in understanding the impact Industry 4.0 will have in an increasingly digital society".*

- *"I feel Government and industry no longer have common goals, have lost any connectiveness the once had, preferring to pursue their own agendas".*

- *"For the last 10 years they have avoided working the tough issues, tech change, climate change and not achieving balance of wages either by gender or cultural diversity or capping the privileged earnings".*

- *"Too much grasp onto legacy industries and attitudes that are being overtaken by new technologies and global issues".*

- *Failure to invest in new technologies, failure to invest in infrastructure that would support new ways of working and living. Continued failure to invest in education at all levels. No genuine leadership*

- *"For the late 30 years we have failed to reform energy, housing, health, education, tax. And under-invested in innovation".*

- *"Government at all levels keep saying that they want to forge close and transparent links with industry, but they seem paralysed to do so by policy, public opinion or red tape".*

- *"No proven track record in keeping pace with the current rate of change".*

- *"I believe that Australian Federal, State and Local Governments should partner with extremely different industries and create shareholder alliances and value across the private sector to drive advances in technological, social, and cultural transformation over the next 10 years and beyond. I find Australia to be lagging in this area in comparison to our counterparts in the Germany, UK, US, Israel and APAC having worked in these jurisdictions during my career tenure".*

- *"Lack of leadership & innovation. A reluctance to take risk or adoption of new technologies. Total lack of understanding of the potential of new economy".*

CHAPTER 9

Australia 2030 predictions

"Just because it is, doesn't mean it should be."
(Australia)

"When you talk to a human in 2035, you'll be talking to someone that's a combination of biological and non-biological intelligence"

~ Ray Kurzweil, futurist, inventor, author and computer scientist

This word cloud represents how Australian professionals feel about technological and scientific developments over the coming decade.

Source: Australia 2030 research Rocky Scopelliti

It is difficult to make predictions, especially about the future. However, there is one prediction we can reliably make — the rate of change will continue to accelerate. It is exemplified by Moore's law, named after Gordon Moore the co-founder of the computer chip company Intel. More than 50 years ago, he observed that the number of transistors on a single die (a measure of computing power) that was available at a fixed price doubled every 18 months or so. The best-known person to make such predictions using an adaption of that principle about the rate of change is Ray Kurzweil. Referred to as the 'restless genius' by *The Wall Street Journal,* Kurzweil has a 30-year track record of accurate predictions. In his research, Kurzweil[164], a world-leading author, computer scientist, inventor, futurist and co-founder of the Singularity University, has made four key observations about the rate of change he describes as *'The Law of Accelerated Change'* in his book titled *'The Age of Spiritual Machines'*[165]. These are:

1. **The Multiplication Effect.** Just like Moore's law theory of price/
    performance of computation (transistor count doubling about every 18
    months), this effect also applies to information technology. Kurzweil
    describes this economic theory as the Law of Accelerating Returns.
2. **Information Enablement.** Once an industry, product or technology
    becomes information enabled, it exhibits its own price/performance
    doubling effect.
3. **Acceleration.** Once doubling begins, it doesn't stop.
4. **Exponential Technologies.** Information technologies are the key
    enabler. Others include as artificial intelligence, robotics, biotech
    and bioinformatics, neuroscience, data science, 3D printing and
    nanotechnology as such technologies.

These observations explain why accelerating technology and evolutionary processes progress in an exponential manner. The term exponential became known to the mass population as health officials, politicians and economists explained the impact of COVID-19 as they fought to 'flatten the curve'.

These accelerating technologies enable the marginal cost of both supply and demand to reduce to virtually zero delivering scale that linear based models of the past, simply cannot achieve. This is the point upon which a traditional value creation system, becomes decoupled from its relationship to growth, supply of scare resources and time. That intersection is where tipping points occur. This effect is what I refer to as the *'new economic physics'* where our traditional notions of value creation through systems of production, distribution and scale are becoming decoupled through dematerialisation, disaggregation and disintermediation. (see Exhibit 9.1).

Exhibit 9.1: A new economic physics — exponential versus linear view of choices

- Exponential systems where the marginal cost of demand/growth or supply/resources/time becomes virtually zero
- Traditional systems of production, distribution and scale are becoming decoupled from growth and supply of scarce resources

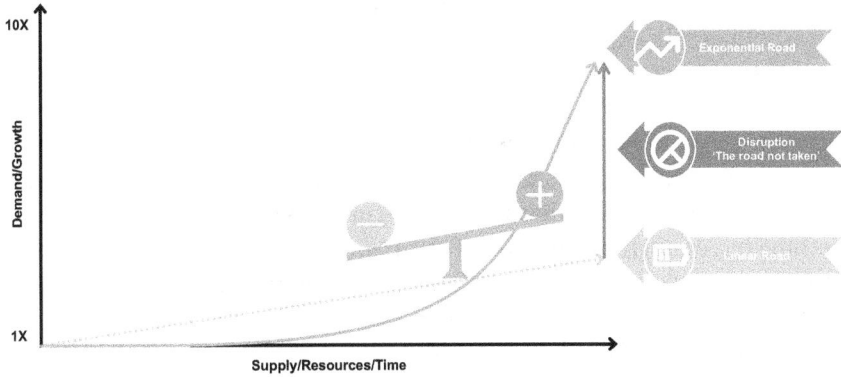

Source: Australia 2030 research Rocky Scopelliti

Before the intersection between the linear and exponential curve is what I refer to as the choices zone. This presents us with a dilemma. Which road do we choose? When making a choice, we are required to make a decision. Viewing a choice as a crossroad, it becomes clear that we must choose one direction or another, but we can't choose both.

It is our curiosity to contemplate the "what if" scenarios about the choice you did not make and hypothesize about the outcomes that might have achieved. After all, that is beauty of 20/20 hindsight. This pondering about the different life we may have lived had we done something differently is central to *'The Road Not Taken.'* But what makes this decade unique, is that for many of the issues before us, be they environmental, economic, regional, trust-related, knowledge-centric, technological or scientific, we will not be able to return to one day and try the 'other' road again. The choices we make around technologies such as artificial intelligence and gene editing are likely to remain with us. This decade will present us with many choices, but for many of those, going backwards will no longer an option. However, like all good mysteries, there are clues along the way. These are tipping points that I've been using throughout each chapter. These tipping points are the moments upon which choices and decisions need to be made. That's what this chapter will explore.

The law of accelerated returns

In his 1999 book *The Age of Spiritual Machines*, Kurzweil proposed "the law of accelerating returns", according to which the rate of change in a wide variety of evolutionary systems (including but not limited to the growth of technologies) tends to increase exponentially. In a subsequent essay in 2001 titled "The Law of Accelerating Returns"[166], Kurzweil argued for extending Moore's law to describe exponential growth of the diverse form of technological progress. Now the key point here is that whenever a technology approaches some kind of a barrier, according to Kurzweil, a new technology will be invented to allow us to cross it. He predicts that such paradigm shifts have and will continue to become increasingly common, leading to "technological change so rapid and profound it represents a rupture in the fabric of human history." He believes the Law of Accelerating Returns implies that a technological singularity will occur before the end of the 21st century, around 2045.

Kurzweil's essay begins:

'An analysis of the history of technology shows that technological change is exponential, contrary to the common-sense 'intuitive linear' view. So, we won't experience 100 years of progress in the 21st century—it will be more like 20,000 years of progress (at today's rate). The 'returns,' such as chip speed and cost-effectiveness, also increase exponentially. There's even exponential growth in the rate of exponential growth. Within a few decades, machine intelligence will surpass human intelligence, leading to the Singularity—technological change so rapid and profound it represents a rupture in the fabric of human history. The implications include the merger of biological and nonbiological intelligence, immortal software-based humans, and ultra-high levels of intelligence that expand outward in the universe at the speed of light.'

2020S

The programmatic decade

To predict change through to 2030 requires an appreciation that the pace of change is accelerating exponentially. In the '90s, Kurzweil made 147 predictions for 2009. In 2010, he reviewed his predictions, 86 per cent of which were correct. He gave himself a "B" grade.

Kurzweil's predictions for this decade, noting his reputation for accuracy, will see significant advances in the areas of nanotechnology, robotics, artificial intelligence (AI). I believe this will be the 'programmatic decade' where technology will allow us to augment our digital, physical, biological and environmental worlds. In other words, we will develop the capacity to control and program evolutionary processes through next generation technologies. Life in 2030 could barely be recognisable if Kurzweil's predictions come true, or at the very least, 86 per cent of them. Let's take a look at how some of these technological and scientific advances may be applied over the coming decade:

The 2020s

- By 2029, computers will have human level intelligence. This decade also marks the revolution in advanced robotics, as an artificial intelligence is expected to pass the Turing test (meaning it can pass for a human being) by the last year of the decade (2029). What follows then will be an era of consolidation in which nonbiological intelligence will undergo exponential growth, eventually leading to the extraordinary expansion contemplated by the Singularity University, in which human intelligence is multiplied by billions by the mid 2040s.
- By the 2030s, nanobots will inhibit our bodies. Nanobots capable of entering the bloodstream to "feed" cells and extract waste will exist (though not necessarily be in wide use) by the end of this decade. They will make the normal mode of human food consumption obsolete. As nanobots flow throughout our bodies, they will keep our bodies healthy and connected to the cloud.
- By the 2030s, virtual reality will be reality. By the later part of this decade, virtual reality will be so high-quality that it will be indistinguishable from real reality. Advancements in sensory and projection technologies will enable us to upload our mind/consciousness by the end of the decade and enter a 3D world where we will see, touch, smell, feel and hear authentically.
- During the 2020s, people will interact with machines in a highly sensory

way just like they do now with other human beings. Keyboards will disappear and verbal communication will be fluid. Computers will have an amazing computing capacity and their software will be able to understand natural language just like human beings. Personal virtual assistants will be widely used.

- During the 2020s, wearable technology will be wireless everywhere. Computers will also be part of our clothing, our home and many other spaces and places. We will finally be able to say goodbye to cable connections, which, in turn, will lead to improvements in technologies such as virtual and augmented reality.
- Before 2030, transportation will be operated by non-human intelligent systems. Autonomous vehicles will include those that use road, rail or air networks that will have the capacity to interact, orchestrate and cooperate with one another.
- By the late 2020s, nanotech-based 3D manufacturing will be in widespread use. We will be 3D printing clothes, buildings and many physical things we have today such as masks that were 3D printed to address the worldwide shortage during the COVID-19 pandemic. This will radically alter the economy as all sorts of products will be produced for a fraction of their traditional manufacture costs and disaggregated from the requirement of labour in the process. The true cost of any product is now the time it takes to download the design schematics and the materials used.
- In the 2020s, we will see the vertical farming revolution. As populations rise and agricultural land shrinks, new ways to produce food will be critical. Vertical farming involves growing produce in nutrient-rich indoor environments and hydroponic plants for fruits and vegetables.

The 4th Industrial Revolution – coming ready or not

Early in this programmatic decade, it is predicted by Kurzweil that we will have the requisite hardware to emulate human intelligence within a $1,000 personal computer, followed shortly by effective software models of human intelligence towards the middle of the decade: this will then be enabled by the continuing exponential growth of brain-scanning technology, which is doubling in bandwidth, temporal and spatial resolution every year, and will be greatly amplified with nanotechnology, allowing us to have a detailed understanding of all the regions of the human brain and to aid in developing human-level machine intelligence by the end of this decade. The revolution in nanotechnology will allow humans to essentially overcome the inherent limitation of biology as we understand it today.

The threat posed by genetically engineered nano pathogens Kurzweil predicts will permanently dissipate by the end of this decade as medical nanobots— infinitely more durable, intelligent and capable than any microorganism— become sufficiently advanced. The many variations of "Human Body 2.0" (as Kurzweil calls it) are incrementally accumulated into this and the following decade, with each organ and body system having its own course of refinement and development. It ultimately consists of a nanotechnological system of nourishment and circulation, obsolescing many internal organs, brain-extension and an improved skeleton.

This will be the decade whereby the augmentation our digital, physical, biological and environmental worlds will see humans becoming hybrids and artificially intelligent. This means that our brains will be able to connect directly to the 'cloud' where computers will augment our existing intelligence. One method for the connection will be made via tiny robots (nanobots) made from DNA strands. This will create a state whereby our thinking will be a hybrid of biological and non-biological thinking.

According to the World Economic Forum, the next decade will bring significant technological and scientific shifts that will alter the speed, scale and impact of development [167].

Exhibit 9.2: Technological and scientific shifts predicted this decade

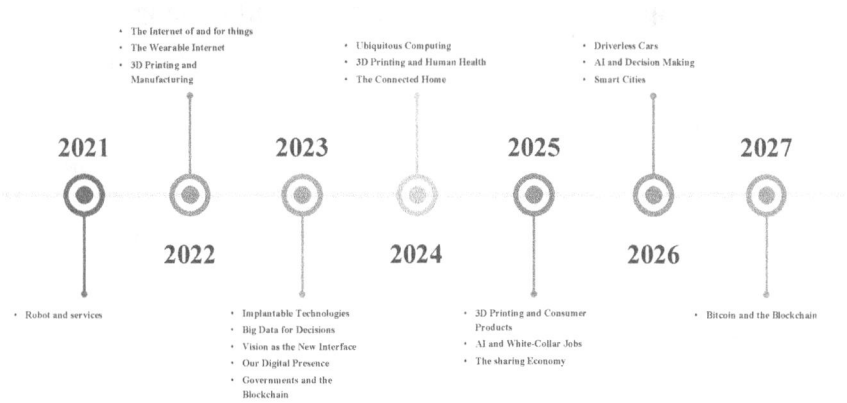

| 2021 | 2022 | 2023 | 2024 | 2025 | 2026 | 2027 |

- The Internet of and for things
- The Wearable Internet
- 3D Printing and Manufacturing

- Ubiquitous Computing
- 3D Printing and Human Health
- The Connected Home

- Driverless Cars
- AI and Decision Making
- Smart Cities

- Robot and services

- Implantable Technologies
- Big Data for Decisions
- Vision as the New Interface
- Our Digital Presence
- Governments and the Blockchain

- 3D Printing and Consumer Products
- AI and White-Collar Jobs
- The sharing Economy

- Bitcoin and the Blockchain

Source: World Economic Forum

Predicting the impact of the 4th Industrial Revolution on Australian professionals in the decade ahead

The Australia 2030 research sought to understand attitudes towards a range of developments over the coming 10 years such as:

1. The speed of technological and scientific change
2. Which technological and scientific developments they believe would have the greatest impact on innovation or disruption in Australia
3. Which technological and scientific developments they believe would have a positive or negative impact on their personal, professional and family life
4. What part of their lives they thought would change the most due to the technological and scientific developments and would we trust robots?
5. Did they think positively or negatively about the role technological and scientific developments would play in their lives
6. Among a range of technological and scientific developments, which ones they predicted they would be experiencing in 2030, and what they felt about them.

Let's take a look at each of these.

1.

The speed of technological and scientific change

As discussed at the beginning of this chapter, however there is one prediction we can make about forthcoming change with absolute certainty: that the rate of change will continue to accelerate and Moore's law and Kurzweil's adaption of the Law of Accelerated Returns say that acceleration will be exponential. It's been widely reported that Australians are early and enthusiastic adopters of technology. The Australia 2030 research overwhelmingly reaffirmed that with 77 per cent of respondents feeling positive about the speed of technological and scientific change over the coming decade and only 8 per cent feeling negative and 15 per cent neutral (see Exhibit 9.3).

Exhibit 9.3: Q. Overall do you feel positive or negative about the speed of technological and scientific change and development over the coming 10 years? (%)

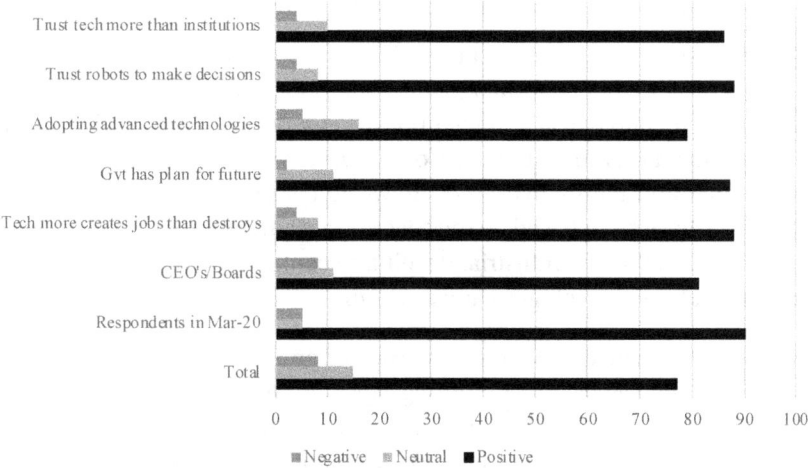

Source: Australia 2030 research Rocky Scopelliti

The emerging Australian professional profile for this decade

The results reveal a profile of those that were more positive and less negative, relative to the total. For example those respondents in March (during the eye of the COVID-19 storm) were 13 per cent more positive as were those who believe that technological and scientific developments create more jobs than they destroy (11 per cent); those who trust automated robots to make decisions on their behalf (11 per cent) and trust technologies more than they trust institutions (9 per cent); those who are confident that the government has effective plans in place (10 per cent), and CEOs and board directors are more positive (3 per cent) about the speed of change relative to other role types.

2.

Which technological and scientific developments do Australian professionals believe would have the greatest impact on innovation or disruption in Australia?

The Australia 2030 research sort to understand how Australian professionals felt about four predictions I am making about the decade ahead. These relate to a range of technological and scientific developments I believe will have the greatest impact on innovation and/or disruption to Australia, our personal, professional and family lives and the world over the next 10 years. These predictions have been based on the speed, scale and impact of developments associated with the 4th Industrial Revolution centred on the augmentation of our digital, physical, biological and environmental worlds. These are:

1. **Our digital world** becomes decentralised and more widely distributed through increasing our computing capacity, automation, connectivity and speed through technologies such as quantum computing; blockchain; Internet-of-Things and 5G mobile networks.

2. **Our physical world** becomes intelligent and interactive such as cities becoming smarter, vehicles becoming autonomous and independently coordinated, factories becoming automated and robotic through technologies such as artificial intelligence, robotics, autonomous vehicles, 3D printing and additive manufacturing, augmented reality, advanced robotics and material science.

3. **Our biological world** becomes blueprinted, programmatic and synthetic. This is achieved by leveraging increased computing power, data intelligence, through artificial intelligence making highly targeted therapies possible, improving quality of life, next generation scientific innovations in a range of new areas such as genetic sequencing and gene editing. Synthetic biology, an emerging area of research, will enable the design and construction of novel artificial biological pathways, organisms or devices, or the redesign of existing natural biological systems, opens up new ways of thinking about medicine and agriculture such as surgical procedures performed remotely, improving agricultural production, precision medicine for treatments through technologies including biotechnologies, neurotechnologies and virtual/augmented reality.

4. **Our environmental world** becomes clean and expands into our universe through renewable energies, colonisation of other planets through technologies including energy capture, storage and transmission, geoengineering and space technologies.

The Australia 2030 research showed overall support for each of those and significantly so for Our Digital World with 48 per cent of respondents who ranked that number 1 as having the greatest impact on innovation and/or disruption in Australia in the next 10 years. In particular, those who believe that technology and scientific developments will impact the world the greatest over the coming decade (60 per cent). Interestingly, during March as the COVID-19 crisis unfolded in Australia, sentiment shifted away from our digital world to our biological world with that being ranked number 1 by 57 per cent of respondents in March 2020 (see Exhibit 9.4).

Exhibit 9.4: Q. Which of the following technological or scientific developments do you believe will have the greatest impact on innovation and/or disruption in Australia in the next 10 years? (%)

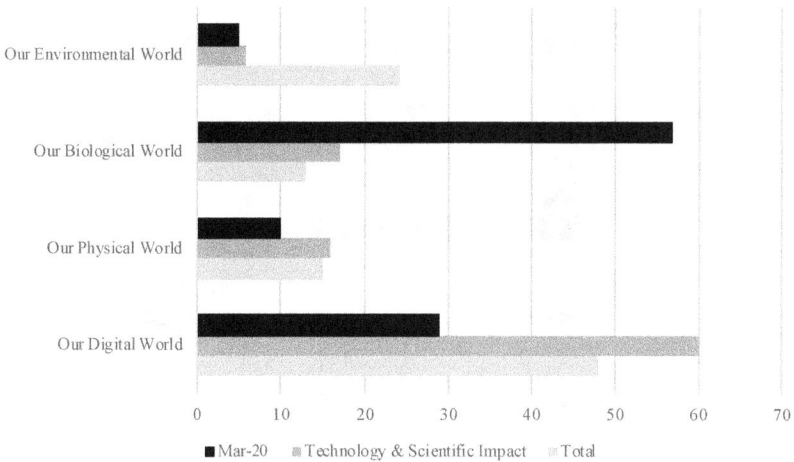

Source: Australia 2030 research Rocky Scopelliti

3.

Which technological and scientific developments do Australian professionals believe would have a positive or negative impact on their personal, professional and family life?

The positive sentiment by Australian professionals in relation to the speed of change shown in Exhibit 9.4, carries through in relation to those four predictions in terms of their impact on their personal, professional and family life over the coming 10 years (see Exhibit 9.5). Interestingly, an uplift across all four developments was observed from those respondents during the month of March 2020 and no generational or demographic difference was observed indicating broad positive effect across Australian professionals.

Exhibit 9.5: Q. Which of the following technological or scientific developments do you believe will have a positive or negative impact on your personal, professional and family life in the next 10 years?

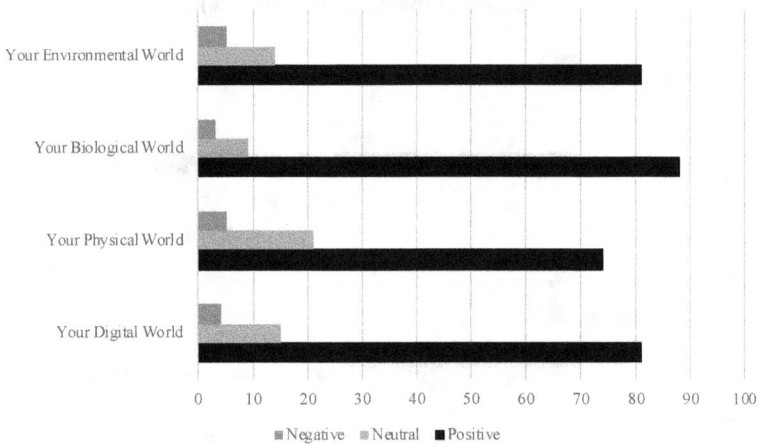

Source: Australia 2030 research Rocky Scopelliti

4.

What part of Australian professionals lives they thought would change the most due to the technological and scientific developments?

While 'job/career' was seen by one in two (47 per cent) Australian professionals to be the part of their life that would change significantly due to technological and scientific developments, 'better physical and mental health' was seen as the least (4 per cent). That quickly changed during March 2020. Sixty-five per cent of respondents in that month chose 'better physical and mental health' to be the part of their life that would change significantly due to technological and scientific developments (see Exhibit 9.6). CEOs and board directors and those who had no confidence that the government has effective plans in place for their future, also chose, but to a lesser extent, 'job/career' as their number one area of impact, 'Better physical and mental health' was their second highest chosen area.

Exhibit 9.6: Q. What part of your life do you think is going to change significantly due to technological and scientific developments in the next 10 years? (%)

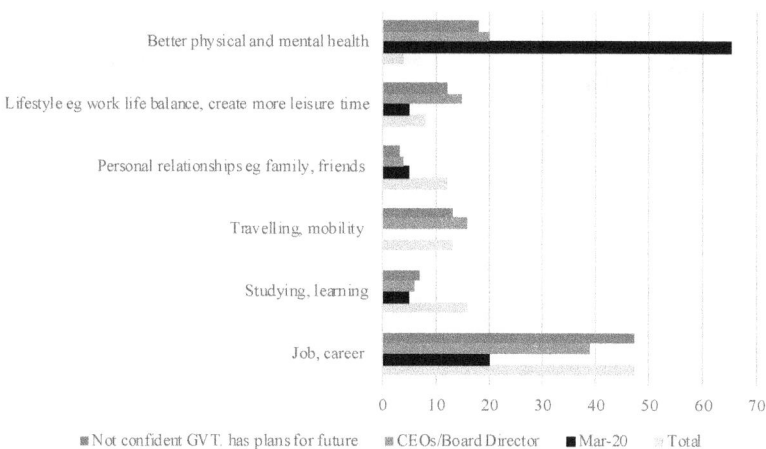

Source: Australia 2030 research Rocky Scopelliti

5.

Among a range of technological and scientific developments, which ones do Australian professionals predict they would be experiencing in 2030, and what they felt about them?

Of all of the four predictions studied in the Australia 2030 research, Australian professionals expect advancements in 'our biological world' (68 per cent) to be what they most likely will experience in 2030 (See Exhibit 9.7). That increased significantly for respondents in March 2020 to 86 per cent. Demographically, more Millennials believe that 'robots will become more intelligent than humans and augmented into their work and family lives (18 per cent).

Exhibit 9.7: Q. Which one of the following statements, do you think you will most likely be experiencing in 2030 (%)?

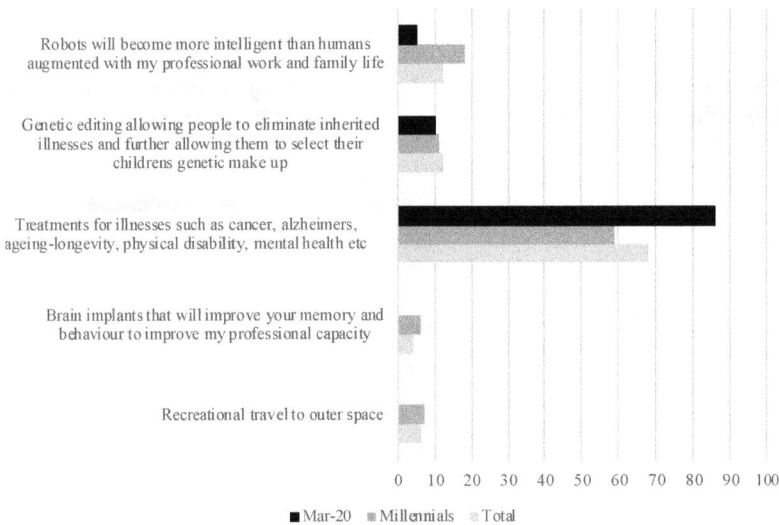

Source: Australia 2030 research Rocky Scopelliti

When it comes to a future retrospective view, Australian professionals overwhelmingly (74 per cent) support the statement that 'technological and scientific advancements have improved our social, cultural and economic future' relative to other statements associated with the pace of change, the rise of Asia and world leadership over the coming decade (see Exhibit 9.8).

Exhibit 9.8: Q. Which one of the following statements, do you think would best describe your perspective in 2030? (%)

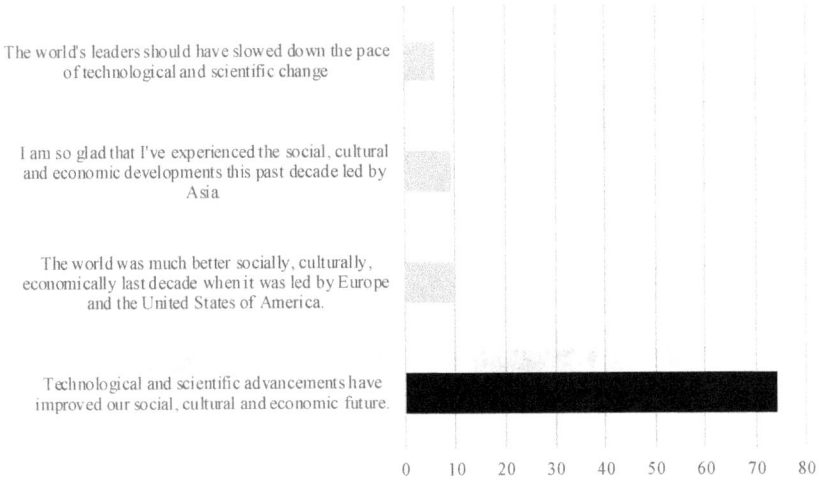

The world's leaders should have slowed down the pace of technological and scientific change

I am so glad that I've experienced the social, cultural and economic developments this past decade led by Asia

The world was much better socially, culturally, economically last decade when it was led by Europe and the United States of America.

Technological and scientific advancements have improved our social, cultural and economic future.

0 10 20 30 40 50 60 70 80

Source: Australia 2030 research Rocky Scopelliti

6.

Did Australian professionals think positively or negatively about the role technological and scientific developments would play in their lives?

Finally, Australian professionals look forward with much optimism over the coming decade about the role that technological and scientific developments will have on their lives with 50 per cent feeling positive and 36 per cent feeling very positive. The Australia 2030 research found no gender or demographic differences with that finding. However, the research found that optimism and positiveness accelerated by respondents in March 2020 to 85 per cent feeling very positive (+ 49 per cent). That suggests in the face of our worst crisis, technological and scientific developments are even more important to the way Australian professionals feel about the coming decade (see Exhibit 9.9).

Exhibit 9.9: Q. When you think about your life over the next 10 years, are you positive or negative about the role of technological and scientific developments? (%)

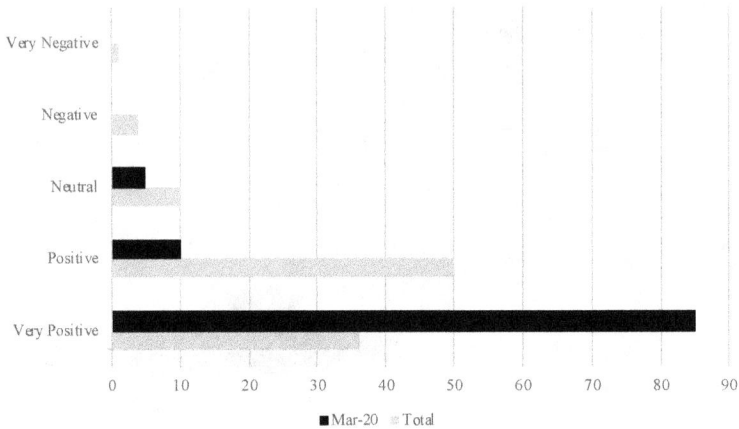

Source: Australia 2030 research Rocky Scopelliti

2030

The journey to the augmentation of our digital, physical, biological and environmental world.

While these megatrends can seem somewhat abstract, we look more closely to practical applications and developments of each as the 'signposts' that we reached the 'tipping points'. In order words, where a series of small technological or scientific developments become significant enough and the speed, scale and impact cross the linear curve and hit the exponential curve. Although the focus of this book is to highlight that we need to be prepared to make decisions at a range of critical crossroads over the next decade, and that for many of those decisions, as a society, we get only one chance, how do the critical decisions related to these particular technology tipping points get made? For some, like gene editing, there are very clear and direct sets of decision. For others its yet to be determined.

The following are just a small drop in the vast ocean of technological and scientific developments underway throughout the world aligned to the major issues discussed throughout this book.

1.

The digital world becomes immersive, decentralised and distributed

Tipping point: *Data becomes hypercritical.*

Data has indeed evolved exponentially doubling roughly every 12 months. According to IDC forecasts[168], by 2025 the global data sphere (defined as the sum of all data created, captured, and replicated on our planet in any given year) is set to reach 163 zettabytes (ten times the 16.1ZB generated in 2016). It describes five key data trends that are expected to change our world.

By 2025, 20 per cent of all data generated will be critical to our daily lives and 10 per cent hypercritical. Twenty-five per cent of that data will be generated in real time, and 95 per cent will originate from a hyper connected Internet of Things world. Cognitive systems (machine learning, natural language processing and artificial intelligence) convert data analysis into intelligence that automates and replicates human actions and judgement at high speed, mass scale, better-than-human quality and lower cost.

Tipping point: *Governments recognise and legislate data as an asset.*

To rebalance the economic, privacy and intellectual property opportunity lost in tax revenue by governments and their citizens to internet giants and other organisations by inadequate data rights, privacy and protections. Regulatory and legislative changes will see citizens regain economic, intellectual property and protection control of their information from their personal, professional and family lives. Some of the early signposts include the General Data Protection Regulation (GDPR) established for the European Union and the Consumer Data Right (CDR) introduced into Australian law. The CDR aims to help you monitor your finances, utilities and other services, and compare and switch between different offerings more easily. The system also aims to encourage innovation and competition between service providers, helping you to access products and services that better suit your specific needs. An example of this is the Australian Farm Data Code aims to promote adoption of digital technology, by ensuring that farmers have comfort in how their data is used, shared and managed. The Australian Farm Data Code (the Code) was developed and adopted by the National Farmers' Federation in consultation with industry[169].

A wonderful friend of mine describes these as relatively early stage responses to "a horse that left the barn quite a while ago". So, the question is, have we already missed the choices zone? Were we ever really given a choice? Is the process of simply opting-in or opting-out adequate consultation? Can data regulation be a fast follower to ways we exploit data, or can we get ahead of the exponential curve – are we making linear responses to an exponential issue? That will be one of the fundamental societal questions we'll face over the next decade.

Why? Because this decade will see our lives becoming exponentially interconnected through more devices, distributed and accessible through the cloud, predictive through personalisation, self-learning through algorithms, self-provisioning through many forms of biometric identification, self-adapting to our ecosystems intelligently, autonomously and in real time. These will be enabled through increased computing capacity, pervasive artificial intelligence and automation. These will in turn, be enabled by pervasive high-speed connectivity, quantum computing, blockchain, the Internet of Things. We need to pay close attention to data regulation because much of life in the coming decade will be data driven.

TIPPING POINT

Augmentation becomes a common tool of lifelong learning.

Our learning systems can no longer keep pace with the speed of change and the exponential growth of knowledge. Yet, as highlighted previously, the need to continuously learn and reskill has never been more important. To meet this need, the way we generate knowledge, curate it and deliver it will become augmented.

Learning programs will be dynamically generated by systems that continuously measure how effectively we are absorbing knowledge. These will be delivered through augmented virtual, augmented and mixed reality. Many new channels will be explored such as immersive holographic delivery and even, potentially in the longer term, direct neural stimulation through implants in our brains. Eventually, our native intelligence will undoubtedly be augmented. For the first time, humans will not be the only, or even primary, originators of the knowledge we consume – techniques such as machine learning will enable the continuous origination of new knowledge without human involvement.

Technological advancements and tipping points referred to earlier in nanotechnology pave the way for implants and neural technologies. These introduce some interesting and fundamental questions on the very nature of 'learning'. If 'knowledge' can simply be directly implanted, or even potentially be externalised and simply accessed on demand from the 'knowledge cloud', what does 'learning' even mean?

Do we need to completely revise the whole concept of learning? What does it mean for the traditional guild mentor models when the availability of knowledge becomes instant? These models underpin not just learning, but organisational hierarchies and operating models. In a world of aging populations as we saw in chapter 5, will neurological knowledge resolve the skills gap issues present in an exponential world? Which of the ASHEN model (Artefacts, Skills, Heuristics, Experience, Natural talent) requirements to execute a job are we comfortable, if any, to have substituted with neurological knowledge?

TIPPING POINT

We can create 3D images indistinguishable from reality to the naked eye.

Holographic technology promises ultra-realistic human presence. The futuristic technology could usher in a new era of ultra-realistic "telepresence" ranging from remote participation in meetings and conferences to virtual on-stage appearances at concerts. Projection and display technology including cameras, will be directly implanted in our eyes and images will be projected directly onto the retina in high resolution and 3D and allow the capture of images. Cochlear implants, developed in Australia, are ubiquitously used to provide auditory communications in both the digital and physical worlds. Three dimensional holographic displays connected to a range of neural implants enhance our visual, auditory perception, interpretation and reasoning.

There have been many significant events this past decade that forced humans to adapt to alternative forms of interactions when physical presence was not an option. For example, the 2010 volcanic eruptions of Eyjafjallajökull in Iceland caused enormous disruption to air travel across western and northern Europe over an initial period of six days in April 2010. The most recent too, COVID-19, that brought the world to a standstill. Each of those events forced behavioural change at scale to the way that we interact. These natural disaster disruptions will present themselves more frequently, as we discussed in chapter 3, as a consequence of climate change. For many, COVID-19 was the first time they used video technologies to go about their daily lives and many may never return to pre-COVID-19 behaviours as they relate to interactions. This will be quite disruptive to the airline and other transport industries and the future of travel.

Advances with 3D imagery where the experience becomes indistinguishable from reality to the naked eye opens up the imagination to consider what other experiences could then be enjoyed when time and space are removed. For example, for those of you who are art lovers, imagine experiencing an afternoon stroll through the Louvre in Paris from the comfort of your lounge room. Or for those of you who love exploring, imagine taking a guided walk through one of the four permanent Australian base stations in the Antarctic, again from the comfort of your lounge room.

TIPPING POINT

Corporate audits, compliance and governance are performed by artificial intelligence on distributed ledger technologies.

Industries, supply chains and ecosystems become programmatic. Services and some service providers become automated and replaced by decentralised autonomous organisations (DAO), also known as decentralised autonomous corporations (DAC). A DAO is a network that runs itself. It's able to operate autonomously due to the clever use of mechanisms that align economic incentives (distribute risks/rewards) using software. The network is structured and incentivised to operate without the use of trust, third parties or centralised powers. In a financial services context, finance becomes 'programmatic'. Distributed ledger technology restructures value chains by disintermediating (removing intermediaries), disaggregating (separating processes into components) and dematerialising (shifting from physical to digital form).

The disruptive impact of a DAO is contained within a property that is often overlooked. That is, it operates in real time. For example, "DOA based audits" would no longer become a point in time retrospective analysis (auditing last year's accounts). Audits occur continuously and in real time. There is significant interest in this model in various areas of government as a primary regulatory innovation. If a regulator knows as soon as there is variation then we can potentially do away with the whole cycle of "record, audit, retrospective fine (and often protracted litigation)" to a mode where regulators intervene with corrective action as soon as variation occurs. Imagine the disruptive impact that would have on the legal, accounting and professional services across regulated markets such as financial services, healthcare, energy etc.

Importantly, imagine the transformative impact this could have on 1st, 2nd and 3rd lines of defence in risk management.

Areas and industries that could be revolutionised by technologies such as quantum computing include:

- *Medicine and materials* – untangling the complexity of molecular and chemical interactions leading to the discovery of new medicines and materials.
- *Supply chain and logistics* – discovering the ideal solution for ultra-efficient logistics and global supply chains.
- *Financial services* – discovering new ways to model financial data and isolating key global risk factors in financial markets.
- *Artificial intelligence* – making aspects of AI such as machine learning more powerful when data sets are very large.

Areas and industries that could be revolutionised by technologies such as blockchain and distributed ledger include:

- *Crypto-currencies* –one of the best-known applications e.g. Bitcoin described as a decentralised virtual currency.
- *Supply chain and logistics* – self executing contracts that transfer risk, value and insurance verifying the transaction or any associated claims with pre and post authorisations and instructions.
- *Currency exchange and remittance* – perhaps the most advanced use cases with financial institutions focused on cross border payments. Pioneers include Ripple
- *Provenance and chain of custody* – time stamping features in agriculture (e.g. from paddock to plate), diamond.
- *Record management* – improve medical record management to store health care data and manage medical records.
- *Decentralised markets* – enabling people anywhere to trade with one another without requiring an institution.

2.

Our physical world becomes interconnected, then intelligent

TIPPING POINT

Global connectivity will be available to connect everyone and everything, everywhere, at ultra-low cost and up to gigabit speeds.

The journey to cheap, ubiquitous, globally pervasive high-speed data connectivity has been an evolution in progress for more than a century. Initially through the mass deployment of telegraph, then telephone networks and then the explosion of widespread data networks that eventually gave birth to the Internet. More recently through deployment around the world of analogue, 2G, 3G and 4G mobile networks and the chequered history of global satellite networks such as Iridium.

There are many signs that this decade will see that vision realised. We have already seen initial commercial deployments of the next generation of cellular networks – 5G – which will deliver unheralded speeds. Global deployment of 5G continues to accelerate. We are seeing mass deployment of networks dedicated to Internet of Things devices including CAT-M, Narrowband-IoT and various low-cost proprietary "Low Power Wide Area Network" (LPWAN) technologies that radically increase coverage for IoT devices. Perhaps most disruptively, we are seeing the deployment of the first Low Earth Orbit (LEO) satellites in massive satellite "constellations" designed to deliver low cost gigabit-level across much of the earth.

This connectivity revolution will enable our cities, buildings and homes to become smarter, more efficient and more liveable. They will underpin the shift of our massive transport systems to autonomous vehicles – making those system not just more efficient, but much safer. They will primary connection between automated and robotic technologies and the artificial intelligence that drives them.

TIPPING POINT

More connected "things" will surpass the number of connected people.

As discussed earlier, as our lives, businesses, cities and appliances, sensors etc become connected and inter-connected into ecosystems, they will generate exponential amounts of data. Analysts predict that by 2025, an average person will interact with connected devices around 4,800 times per day (one interaction every 18 seconds)[170]. Most estimates say the number of connected things doubled the number of connected people in 2018-19. Gartner predict about 20 billion devices in the next year or so – so that will be 3-4 times the number of connected people). Cisco is predicting up to 500B connected devices by 2030171 and it's a reasonable bet to make that there will be an order of magnitude more things connected than people sometime this decade.

Connection will be between devices, sensors, appliances, equipment, products, services, locations and people in our homes, transport, clothes, accessories, cities, energy and production and much more. Importantly, these will become inter-connected ecosystems. Connectivity has, and will continue to, fundamentally change the way we manage our lives, businesses, industries and governments. Supply chains will become more efficient, assets better utilised, and healthcare better managed through monitoring systems, transactions processed in 'trusted' real time, crime prevented. Breakthroughs in material science, locomotion and sensory technologies have improved the orchestration of networked machines and their ability to move in different environments. Cognition enables them to be choreographed.

TIPPING POINT

Machine natural language abilities in understanding, interacting, timing and speaking surpass those of humans.

Virtual assistants will become personalised and extend beyond our smartphones that are autonomous, connected, electric and cognitive. Much of the activity revolves around automotive technologies known by the acronym ACES—vehicles that are autonomous, connected, electric, and shared. Some signposts of technological advancements in this area include digital translators that are now as accurate as human translators and digital transcription, which is now more accurate than skilled human transcription.

TIPPING POINT

More mobility through using autonomous vehicles.

Autonomous-car technologies will soon transform what "riding" means. Autonomy, expressed along a framework from the Society of Automotive Engineers ranging from level 0 (full driver control at all times) to level 5 (the full-time performance by an automated driving system of all aspects of driving under all roadway conditions), is being pursued. Automakers predict that driverless cars will account for 10 per cent of all cars in the US by 2026.

The question is who will hold moral accountability for the decisions that automated vehicles make?

TIPPING POINT

We will be forced to more aggressively confront how we produce in the face of growing resource constraint. Circular economies will become the norm.

To keep volumes of major commodities such as metals in line with economic growth, we will need to more quickly embrace circular models: sourcing much less from virgin materials, using recycled content and remanufactured products, and generally rethinking the material economy. Water will be a stressed resource and it seems likely that many cities will be in a constant state of water shortage. We will need more investment in water tech and desalination to help.

We'll also change how we produce new materials. Much new materials engineering is now effectively done in software rather than the lab. We are reaching the point where software has the ability to take a set of requirements and constraints and design a material to meet those. For example, this is how a lot of exploration for new high temperature superconductors and some classes of pharmaceuticals is done. Work is well underway on the next phase of this journey, where the software also designs the production process for that material.

New materials are emerging that are lighter, stronger, recyclable, conductive, transparent and adaptive. For example: bio-inspired plastic that is light enough to permit flight, thin enough to accommodate flexibility; ultrathin platinum–hydrogen fuel cell vehicles that could provide clean transportation in the future; graphene, with its strong, conductive, flexible properties, which could be applied to next generation electronic devices or even sewn into our clothing; and stanene, which was created as a topological insulator and scientists say is the natural successor to copper interconnects in computers.

TIPPING POINT

The 'instant economy' is created through on demand production and delivery.

3D printing and additive manufacturing cover a range of processes and technologies that produce physical objects layer by layer in an additive manner from a digital 3D model. According to an ABC News report172, this development could have a bigger impact on economies and society than the internet. Industry analysts believe we are only at the very beginning of understanding the potential of these technologies. Originally, 3D printing was used for rapid prototyping as a means of fast tracking and reducing the costs of product development but is now being considered for much broader application.

These technologies allow for new ways of distributed, customisable manufacturing.

For example, the technology has the potential to bring the production process closer to its destination, or directly to its customer – eliminating the logistics and supply chain costs associated with mass production. The distributed nature of the technology reduces production risks with smaller batch runs. These have significant impacts on global economies through shifting from production and distribution models, to more localised on-demand models, which will reduce imbalances in imports and exports between developing and developed nations.

Some of the industries benefiting from or disrupted by this 3D printing technology include:

- *Medical and dental* – customisation and personalisation to meet medical standards for items such as hip and knee implants, dental crowns, and prosthetics. Although possibly decades away, the technology is being researched and developed for printing skin, bone, tissue, pharmaceuticals and human organs.
- *Aerospace* – companies such as GE/Morris Technologies, Airbus/EADS, Rolls-Royce, BAE Systems and Boeing are using the technology in their research and developments and production. Many of the injectors for jet engines are now 3D printed because they can be made with tighter tolerances than standard casting and milling approaches.
- *Automotive* – prototyping high-performance equipment in motor racing through to prototyping for mass-produced vehicles. This is evolving to the production of on-demand spare/replacement parts, as opposed to holding inventory.
- *Jewellery* – reducing the reliance on high levels of expertise in the manufacturing process such as fabrication, mould making, casting, electroplating, forging, silver/goldsmithing, stone cutting, engraving and polishing.
- *Architecture* – prototyping models of building or engineering designs with some visionary architects looking at direct construction methods.

Areas and industries that could be revolutionised by combining technologies such as robotics, artificial intelligence, blockchain and distributed ledger include:

- *Manufacturing* – assembly lines e.g. automotive industry
- *Mining* – extracting and processing minerals from the ground
- *Healthcare* – surgical procedures
- *Agriculture* – weed and pesticide management
- *Entertainment* – science fiction motion pictures.

3.

Our biological world becomes augmented with implantable technologies

Tipping point: *The rise of synthetic biology from our understanding of what genes do.*

What's driving the shift is the move from merely reading genomes to writing them. The Human Genome Project enabled us to map genes which was one of the required precursors to progressively learning what specific genes actually do. For example, if we know which gene is actually causing a cancer, we can deploy therapies for that particular mutation rather than merely classifying the tumour based on where it is located like the breast or the prostate gland. That's been helpful, but somewhat limited. Actually, being able to write genes is something else entirely. Think about a genetic disease like cystic fibrosis. By replacing the mutated gene with a healthy one, we can cure diseases that afflict millions.

We will be forced to confront some very fundamental questions as this capability unfolds. What is the boundary between negative mutation and natural variation? Mutation plays an important role in natural selection – Darwinian theory would hold that without variation, evolution ceases. We will be forced to confront questions of whether the expression of a mutation is a condition, disease or disability rather than just a characteristic. For example, how do you feel about paving the way for 'designer' children? For fundamentally, this crossroad questions who we actually are as organisms, and critically, it will not be a road from which we could ever change course.

TIPPING POINT

The first implantable mobile phone.

We are becoming connected to more and more devices which are increasingly becoming connected to our bodies. These devices are not just being worn, but are also being implanted into our bodies, communicating, providing location and monitoring our behaviour and health function. Some of the enabling technologies include procedures performed remotely, improved agricultural production, precision medicine for treatments, bionic organs through technologies including biotechnologies, neurotechnologies, nanotechnologies and virtual/augmented reality.

TIPPING POINT

The average health span will increase by the convergence of genome sequencing, CRISPR technologies, AI, quantum computing, and cellular medicine.

Technology breakthroughs over the past 15 years have significantly reduced scientific research and development costs. Along with increased computing power, data intelligence through artificial intelligence will make highly targeted therapies possible, improving quality of life. These breakthroughs have paved the way for next generation scientific innovations in a range of new areas such as genetic sequencing and gene editing. Synthetic biology, an emerging area of research that enables the design and construction of novel artificial biological pathways, organisms or devices, or the redesign of existing natural biological systems, opens up new ways of thinking about medicine and agriculture.

TIPPING POINT

The first transplant of a complete 3D printed body organ.

3D printing and additive manufacturing will allow customisation and personalisation to meet medical and dental standards for items such as hip and knee implants, dental crowns and prosthetics. Doctor or experts on demand will distribute expertise from specialist health care, to field services technicians, to advisory or professional services including the capacity to perform remote surgery and patient monitoring.

Although replacement of complete organs with 3D printed versions may possibly be decades away, the technology is already being researched for printing significant body components such as bone, fat tissue and muscle tissue. Parts of organs such as skin and small sections of heart tissue can already be printed and are being used therapeutically today.

This technology already does bring up some interesting ethical and legal issues. For example, legal and regulatory battel are already being fought over who, if anyone, own a map of your genome. Similarly, any 3D printed replica of your existing organs must follow a blueprint. Who should naturally own the creative rights to a blueprint of YOUR organs?

TIPPING POINT

Quantum computing becomes available in the cloud as a service.

As far as has been publicly disclosed, private investment in quantum computing is growing rapidly. There is, however, concern that a large part is going into quantum software – despite the fact there is little clarity on which hardware architecture and approaches will become dominant.

Today's quantum computers need to operate at very low temperatures — until recently, temperatures that challenge today's engineering capability. Recent developments have seen temperatures rise by almost 3 orders of magnitude – but only to about 1.5 Kelvin – half the temperature of the coldest interstellar space!

Some of the impacts are for example: the collapse of most forms of cryptography used today; the use of quantum computing to simulate quantum-level physics and chemistry will radically accelerate development of new purpose-designed materials and chemicals; radically reducing the time required for all manner of optimisation tasks – potentially allowing for continuous, real-time optimisation of all manner of technical, business and lifestyle processes. We already see that machine developed optimisation strategies for many processes surpass human capability to develop and execute them – we'll likely see many surpassing our ability to even understand how they are being optimised.

TIPPING POINT

Cellular agriculture moves from the lab into inner cities, providing high-quality protein that is cheaper and healthier.

This next decade will witness the birth of the most ethical, nutritious, and environmentally sustainable protein production system devised by humankind. For example, cell replication/reproduction technologies 'cellular agriculture' will allow the production of beef, chicken, and fish anywhere, on-demand, with far higher nutritional content, and a vastly lower environmental footprint than traditional livestock options. Precision nutrition which is the combination of instrumenting the human body, AI-based analytics and the ability to custom make foods mean we can tailor the right food for my system at this point in time. This will potentially eliminate a whole raft of dietary-based conditions and other conditions which can be treated by dietary means. This megatrend is enabled by the convergence of biotechnology, materials science, machine learning, and Agtech.

However, there are also some very significant ethical, social and legal questions that need to be addressed and regulated. These will challenge our notion of what it means to be human, what information about our bodies, such as DNA, should be shared, and what rights people have on genetics and their modification. The ability to edit and modify biology (plant, animal and human organisms) provides new ways of thinking about food production, disease management and medicine. While academics, scientists, institutions, governments and communities work through those issues, we will explore the possibilities and advances in biotechnologies, neurotechnologies and brain science and how they will operate within our own biology and change how we interface with the world.

There are three main industries where biotechnology offers major possibilities: medical healthcare; agricultural crop production; and non-food uses of crops such as biofuels, biomaterials, and vegetable oils. There is also engineering organisms to produce particular materials or destroy particular waste. Examples of applications include:

- *Medical healthcare* – precision medicine that relies upon complex data on the patient's molecular make-up and genomic, transcriptomic, proteomic, metabolic and microbiomic profiles to deliver customised therapies.
- *Agriculture* – food security to meet the global demands for food, its quality *and health care. Gene editing offers precision crop improvement.*
- *Bio refineries* – creating biofuels using renewable feedstocks that exploit catalytic properties of microorganisms and repurpose CO_2.

Neurotechnologies help us understand the brain and how to influence consciousness, thought and higher order brain activities. While developments with brain imaging have revolutionised the field, other developments include decoding what we are thinking through to new chemicals that influence our brain. New technologies are enabling the measurement, analysis, translation and visualisation of the chemical and electrical signals that exist in the brain.

They offer opportunities to address current neurological conditions and physical disabilities. It is anticipated that these technologies will lead to new medical breakthroughs in areas such as:

- *Read and write to the brain* – the ability to correct deficiencies or add enhancements such as thought access and influence.
- *Enabling new forms of cognitive computing* – increased knowledge on brain functioning could improve the design of machine learning algorithms.
- *Perception* – influencing the brain in more precise ways such as sense of self, our experiences and what constitutes reality.

Those of us that saw the movie "Eternal Sunshine of the Spotless Mind" about the ability to use the same technology to write or erase memories and perhaps even behaviours and personality traits, invites some very important questions. Such as, if aspects of memory, thinking patterns, emotional responses etc are rewritten or edited, where is the boundary where we cease to be us?

4.

Our environmental world becomes cleaner and extended into space

TIPPING POINT

Renewable energy costs become cheaper than fossil fuel production.

The price of solar energy has plummeted over the last decade by more than 90 per cent and, as the World Economic Forum reports, wind and solar now produce cheaper energy than coal and gas in North America. Solar-power plants now cost less than 10 per cent to construct today than they did only seven years ago. We can only expect renewable energy to get cheaper in the future as a revolution in materials science allows us to build more efficient solar panels and wind turbines. That will mean that we can not only benefit from cleaner energy, we will also be able to get it cheaper than fossil fuels used today. If the trend continues, solar and wind energy could grow from 4 per cent of power generation today, to 36 per cent of global electricity supply by 2035[173]. For example, in Queensland today, the spot energy price often goes negative during the day because of solar produced energy.

TIPPING POINT

We see the emergence of an "Energy storage as a Service" business model.

There is already significant progress and we're likely to see much greater progress in commercial and household application for energy storage solutions. The energy industry is now entering another industrial revolution with advances in clean energy and storage and distribution technologies and the global momentum to bring them into commercial operations. As well as the economic benefits of reduced energy costs, most countries recognise the environmental benefits associated and are reforming policies to support their expansion into consumer and industrial markets.

Renewable generation already exceeds fossil/nuclear generation from time to time in various countries. But storage for distributed renewable energy production is still prohibitively capital intensive in most applications. Large scale storage facilities (typically Lithium battery banks today) are currently mainly used for short term grid stabilisation. New technologies such as bulk hydrogen storage would enable broad scale "Storage as a Service" offerings.

TIPPING POINT

Device, vehicle, building systems-to-grid regulatory barriers lifted.

Stationary storage technologies will radically unlock the potential of transient renewable generation technologies like wind and solar. Mobile battery technologies on the other hand will significantly increase energy density, cycle life, charge and discharge rate, charge/discharge losses and self-discharge rate. These will enable a whole raft of new applications (long range EVs being just the first). Electric planes and a plethora of devices requiring wired power or fossil fuel engines will progressively move to long term battery power. Improved battery technology will also radically accelerate the diffusion of the Internet-of-things.

E-mobility is expanding, and we now see electric cars, bikes, scooters and kick bikes in the streets. Utilisation of all of these vehicles will be extended beyond their intended use as means of transport to also include energy storage: they will charge when renewable energy is abundant in the system and feed energy back into micro-grid's battery when needed.

TIPPING POINT

Sustained exploration of the Moon and Mars.

Space technologies are developed for use in spaceflight, satellites, or space exploration including spacecraft, space stations and supporting equipment and infrastructure. Technology areas such as cryogenic propellant production and management, sustainable energy generation, storage and distribution, efficient and affordable propulsion systems, autonomous operations, rover mobility, and advanced avionics are expected this decade for sustained exploration.

Commercial developments such as SpaceX have significantly reduced the launch costs. Other developments are exploring enabling aircraft to fly in low earth orbit without landing strips and facilities. Richard Branson has also been looking at space tourism. Along with better access to space, new industries will emerge to support this rapidly developing ecosystem. New advanced materials detailed earlier are at the forefront of supporting the design of craft and humans in space. Advancements in computing, advanced robotics, artificial intelligence and renewable energy will all play a significant role in improving the economics of space technologies.

More than 70 countries have owned or operated a satellite in orbit with more plans afoot. There are an estimated 2,666 satellites in orbit today and approximately 2,600 decommissioned satellites floating in space. Space X has announced plans to launch a constellation of up to 42,000 satellites in orbit this coming decade[174]. The problem is the impact that these launches are likely to have on ground-based astronomical observations. Astrophysicists carrying out long exposures in the order of 15 seconds already have images corrupted by passing satellite trails many times brighter than the faint galaxies they are trying to study. However, the launch of enormous numbers of new satellites will exacerbate this effect; transforming the sky above our collective heads into a seething mass of moving objects that will permanently alter our environment.

KEY POINTS

- The 4th Industrial Revolution will see many accelerating technologies and evolutionary processes that will progress in an exponential manner. These accelerating technologies enable the marginal cost of both supply and demand to reduce to virtually zero delivering scale that linear based models of the past simply cannot achieve. This is the point upon which a traditional value creation model becomes decoupled from growth and the supply of scare resources and time. This is the new economic physics.

- The revolution in nanotechnology will allow humans to overcome the inherent limitations of biology, as no matter how much we finetune our biology, we will never be as capable otherwise. Early in this decade, we will have the requisite hardware to emulate human intelligence within a $1,000 personal computer, followed shortly by effective software models of human intelligence towards the middle of the decade: this will be enabled by the continuing exponential growth of brain-scanning technology, which is doubling in bandwidth, temporal and spatial resolution every year, and will be greatly amplified with nanotechnology, allowing us to have a detailed understanding of all the regions of the human brain and to aid in developing human-level machine intelligence by the end of this decade.
- This will be the decade whereby the augmentation of our digital, physical, biological and environmental worlds will see humans becoming hybrids and artificially intelligent. This means that our brains will be able to connect directly to the 'cloud' where computers will augment our existing intelligence. The connection will be made via tiny robots (nanobots) made from DNA strands. This will create a state whereby our thinking will be a hybrid of biological and non-biological thinking.
- It's been widely reported that Australians are early and enthusiastic adopters of technology. Looking to the decade ahead, the Australia 2030 research overwhelmingly reaffirmed that with 77 per cent of respondents feeling positive about the speed of technological and scientific change over the coming decade and only 8 per cent feeling negative and 15 per cent neutral.
- The Australia 2030 research showed overall support for each of our predications based on our digital, physical, biological and digital worlds. Significantly so for our digital world with 48 per cent of respondents who believe that will have the greatest impact on innovation and/or disruption in Australia in the next 10 years. Interestingly, during March 2020 as the COVID-19 crisis unfolded in Australia, sentiment shifted away from our digital world to our biological world by 57 per cent of respondents.
- When it comes to a future (2030) retrospective view, Australian professionals overwhelmingly (74 per cent) support the statement that 'technological and scientific advancements have improved our social, cultural and economic future' relative to other statements associated with the pace of change, the rise of Asia and world leadership over the coming decade.
- The positive sentiment by Australian professionals in relation to the speed of change carries through in relation to those four predictions (digital, physical, biological and environmental) in term of their positive impact on their personal, professional and family life over the coming 10 years.

- Of all of the four predictions studied in the Australia 2030 research, Australian professionals expect advancements in 'our biological world' (68 per cent) to be the most likely they will experience in 2030. That increased significantly for respondents in March 2020 during the COVID-19 crisis to 86 per cent.
- Finally, Australian professionals look forward with much optimism over the coming decade about the role that technological and scientific developments will have on their lives with 50 per cent feeling positive and 36 per cent feeling very positive. The Australia 2030 research found no gender, demographic differences with that finding. However, the research found that optimism and positiveness accelerated by respondents in March 2020 to 85 per cent feeling very positive. That suggests in the face of our worst crisis, technological and scientific developments become even more important to the way Australian professionals feel about the coming decade.

The technological and scientific decade ahead

Most Australian professionals are optimistic, enthusiastic and ready to embrace the technological and scientific developments and the impact of the digital, physical, biological and environmental worlds to their lives. This word cloud represents that enthusiasm.

Source: Australia 2030 research Rocky Scopelliti

I'll close this chapter by sharing some of the qualitative quotes from the 133 respondents who provided explanations for their choice.

IN AUSTRALIAN PROFESSIONALS' OWN WORDS

- *"I like technology and I'm an early adopter..."*

- *"More developments are crucial, even if there is short term pain to implement them."*

- *"Until we connect technology and economic growth with a new economy that does not openly destroy the world, we will have limited success."*

- *"Technology innovation will allow people to do things faster, better and accurately solving significant problems currently facing the world. It will however increase competition and create areas of unemployment and disadvantage as wealth concentration continues to accelerate."*

- *"Very excited about adopting and experiencing new developments."*

- *"The net benefit in the last 10 years has been positive and I see no reason that this will change in the next 10 years."*

- *"At the end of the day – you don't know what you don't know. Technology has the ability to make some great changes, but when used the wrong way has a detrimental impact. My main concern is Social Media and its impact on our youth. From early access to platforms before minds are developed enough to deal with the emotions and chemical reaction, to parents that are so absorbed by Social Media that they don't even interact with the children on outings. Personally, I hate social media and raising teenagers in today's world is terrifying! IMHO social media may impact the world more the climate change."*

- *"One of the greatest times to be alive at the threshold of massive technological developments, reasonably stable Governments and international institutions."*

- *"Overall I am optimistic about the impact of technological and scientific developments. My caution is the application and governance."*

- *"I think it will only create more opportunities as long as I am prepared to up-skill myself and be prepared to work with younger generations and accept them as my peers and potentially as my manager."*

- *"Our ability to solve complex problems has improved and technology is forcing us to think about our own thought processes & how our brain works. You can't create intelligent thinking until you understand what you are creating."*

- *"Technology already has the ability to significantly impact on our lives. the critical issue is will (or rather how) we let it......."*

- *"I believe humanity will find a way to balance out risks and advantages of new technologies."*

- *"We have many talented researchers in Australia leading the way in this area."*

- *"While I think technological developments will have a positive impact on quality of life generally, my concern is career and employment for my children, without employment they won't be in a position to use the technology."*

- *"Scientific developments will help health advancements primarily and this can only be encouraging, notwithstanding this technology will improve infrastructure and businesses"*

- *"Harnessing science and technology – data driven rationale – is the best outcome when administered in the interests of the broader communities that people live in."*

- *"Scientific endeavour is essential to environmental and corporate sustainability."*

- *"In my experience of work, study, medical requirements, and keeping in touch with family, friends and colleagues have all been made easier and quicker due to technological and scientific developments. One big positive is the way technology has helped us with the recent bush fires in my area."*

- *"Mixed, some positives personally and professionally because those with good intentions will use tech for benefit of society and a lot of good will come with these developments. Equally, personally I see negatives because most of the groups we should trust are all compromised due to the financial benefit they receive to implement new policies and services. The commercial sector is not made for the benefit of society, but for the benefit of shareholders, so professionally we as a business (the one I work for) that supports the commercial sector will continue to abuse the power that money and anonymity affords our clients to gain more control and market to the masses whatever they will buy whether it is a true benefit or not."*

- *"I think advances will create new possibilities we can't even yet imagine".*

- *"Misunderstanding and misuse of neural networks, deep learning and big data. Abuse of privacy and technology infrastructure by Government and criminal through cyber with humanity utterly unprepared. Psychological negative effects of technology (digital addiction)."*

- *"I'm a tech evangelist and a research scientist (Degree in Science many years ago - I believe that continual innovation, and acceptance of change, is how the world has evolved - and should continue to evolve (even if it makes us uncomfortable in the short term)."*

- *"I am fortunate to be in a position to be part of the change. I am very aware that not everyone is capable and ready for this change and I can see lots of people around me who share this view so I am hopeful that we will shape the future which is more inclusive and fairer than where our current systems have ended up."*

-

Conclusion

"To this day, I am delighted with my choice [on moving to Australia] and firmly believe that there is no better place to undertake research. Australia offers a culture of academic freedom, openness to ideas, and an amazing willingness to pursue goals that are ambitious. And the results speak for themselves - we have achieved tremendous success in our endeavour, largely because we gave things a go that the rest of the world didn't dare to try."[175]

~ Professor Michelle Simmons – 2017 Australian of the year extract from her speech. Australian Research Council Laureate Fellow and Director of the Centre of Excellence for Quantum Computation and Communication Technology at UNSW

In Professor Michelle Simmons' 2017 Australia Day speech she pointed out that Australia is a country of great spirit, enormous promise and she felt that this was something outsiders don't always appreciate. But she also observed that 'Australians are natural discoverers with inherent scepticism towards dogma, an openness, collaborative spirit and problem-solvers who like to get things done'. That observation is why I felt it is so important to acknowledge the technological, scientific and environmental developments by the traditional owners of the land and the contribution made by them as the world's oldest continuous living culture. They've employed scientific and technological methods of data collection, such as observation and experimentation, for tens of thousands of years and understand the intricate and inextricable interconnection between the physical, chemical and biological sciences and the social sciences more widely.

That heritage, together with Australia's modern history, ingenuity and innovativeness, gives us permission and the confidence to play a significant role in the world's future development of the 4th Industrial Revolution. The

Australia 2030 research reaffirmed that belief. But to do so, requires systems leadership that's been absent in the past decade in the relationships between government, industry, academic and research institutions, and importantly the spirit of Australian citizens. The past decade closed with an example of the absence of systems leadership in a crisis, but then we began this new decade with another crisis where we are all seeing and experiencing how truly remarkable and effective responses can be when systems leadership does manifest.

So, for the coming decade, which road will we take? Will it be a regression to the politics of the past decade, exemplified by expedience and a lack of genuine leadership on major societal issues? Will we simply maintain a holding pattern until the next clear and present crisis emerges – with no time then for us to consider who we want to be, and hence how we want to respond? Or will we strive for entrenched systems leadership, actively considering and debating the future we desire and our place in the world? Will we take an active role in taking the new worlds that today only exist in our imaginations and collectively pursue strategies to create them?

Before we can address those questions, we must first ask 'where the bloody hell are we politically, economically, environmentally, regionally, socially, trustworthily, ingeniously, scientifically and technologically?' What is clear from the Australia 2030 research is that Australian professionals overwhelmingly believe that individual citizens have the greatest role to play in making the world, and Australia a better place over the coming 10 years.

I hope this book has encouraged you to get your juvenescence on. Importantly, I hope the insights in this book inspire you to decide which road to take to create the social, cultural, economic and technological future you desire. Regardless of the road you do take, I hope it encourages you to be a driver on it, rather than simply a passenger.

ABOUT THE AUTHOR

Rocky Scopelliti

Futurologist & Author

Rocky Scopelliti is a world-renowned futurologist. His pioneering behavioural economics research on the confluence of demographic change associated with Millennials and digital technology has influenced the way we think about our social, cultural, economic and technological future. His last book *Youthquake 4.0 - A Whole Generation and The New Industrial Revolution* has now being translated into Vietnamese, Korean, Indonesian and Chinese languages.

As a media commentator, his unique insights have featured on SKY Business News, The Australian Financial Review, ABC Radio National, The Economist, Forbes and Bloomberg. As an international keynote speaker, his presentations have captivated audiences across Asia Pacific, the USA and Europe including Mobile World Congress. As a thought leader, more than150 boards and leadership teams, including Fortune 100 corporations, each year seek his advice on strategy.

In an executive capacity, he is a member of the Optus Business Leadership team as the director, Centre for Industry 4.0 where he leads the specialist team creating world-class thought leadership and innovation on the 4th Industrial Revolution.

In a non-executive capacity, he is a director on the board of Community First Credit Union and on the technology advisory board for REST Super, a member of the Australian Payments Council and on the advisory boards of SP Jain School of Global Management and Wake by Reach.

Educated in Australia and trained in the USA at Sydney and Stanford Universities respectively, he has a Graduate Diploma in Corporate Management and an MBA. He is also a graduate and member of the Australian Institute of Company Directors.

REFERENCES

1   Hugh Mackay, (2018), *'Australia Reimagined.'* (Macmillan Publishing Australia)

2   https://www.afr.com/politics/federal/pm-s-17b-stimulus-to-avoid-recession-20200311-p548ub

3   Australian Government Treasury, (March 2020), *'Economic Response to the Coronavirus'*

4   ABC News, (May 2020), *'JobKeeper and the coronavirus stimulus bill has been slashed, but what can we do with the extra cash?*

5   Visual Capitalist, (May 2020), *'Zoom is Now Worth More Than the World's 7 Biggest Airlines'*

6   Australian Government Parliamentary Budget Office, (February 2019), *'Australia's Ageing Population – Understanding the fiscal impacts over the next decade' Report No. 02/2019'*

7   World Economic Forum, (September 2019), *'Systems leadership can change the world – but what exactly is it?'*

8   Klaus Schwab, (2017) (2016), *"The Fourth Industrial Revolution"* Crown Publishing Group

9   Ray Kurzweil, (2006), *'The Singularity is Near'* (Penguin Publishing)

10 Ray Kurzweil, (March 2001), *'The Law of Accelerated Returns'*

11 Charlie Magee (1993), *"The Age of Imagination: Coming Soon to a Civilization Near You"*

12 https://www.theaustralian.com.au/in-depth/full-transcript-of-kevin-rudds-farewell-speech/news-story/a4d563954f69760c6277c6836fc07f51

13 Professor Ian McAllister, Australian National University, (December 2019), *'Trust in Government hits all time low'*

14 Steven Scott, Courier Mail (July 2013), Prime Minister Kevin Rudd takes power to change leaders off faceless men and hands it to Labor rank-and-file'.

15 https://www.heraldsun.com.au/blogs/andrew-bolt/the-plot/news-story/5b21b5d02418893fed9c09dda08c3de1

16 ABC News Australia. Emma Rodgers, (22 April 2010), *"Combet defends insulation backflip".*

17 The Canberra Times. Nick Toscano; Ben Schneiders, (19 August 2015). *"Health Services Union former leader Kathy Jackson ordered to repay $1.4m".*

18 ABC News. Elizabeth Byrne (26 February 2015). *"Peter Slipper Cabcharge case: Former speaker wins appeal against dishonesty charges".* ABC News; Slipper v Turner [2015] ACTSC 27; Inman, Michael. "Peter Slipper conviction overturned by Canberra court".

19   https://www.abc.net.au/news/2020-02-18/obeid-trial-reconstructed-fiction-defence-argues/11975332

20   Australian Human Rights Commission, (2014), *'The Forgotten Children'.* National Inquiry into Children in Immigration Detention.

21   ABC News, (April 2016), *'PNG's Supreme Court rules detention of asylum seekers on Manus Island is illegal'*

22   The Australian, (April 2017), *'No asylum seeker on Manus Island, Nauru will come to Australia'*

23   The Parliament of the Commonwealth of Australia. House of Representatives (13 February 2019). *"Home Affairs Legislation Amendment (Miscellaneous Measures) Bill 2019. No., 2019A: Bill for an Act to amend the law relating to migration, customs and passenger movement charge, and for related purposes (as passed by both houses)"* (PDF)

24   The Conversation. Michelle Grattan, (16 April 2014), *"Barry O'Farrell quits as NSW Premier over ICAC 'memory fail'".*

25   ABC News. Australia. 16 April 2014, *"NSW Premier Barry O'Farrell to resign over evidence he gave to ICAC".*

26   http://www.news.com.au/finance/work/leaders/former-nsw-premier-barry-ofarrell-cleared-of-any-wrongdoing-in-icac-report/news-story/47241d2d611119338d6ce0d177371028 |Date=3 August 2017

27   Adam Gartrell, (19 July 2015), *"Bishop's press conference is more evidence she's unfit for the Speaker's chair".* Sydney Morning Herald.

28   The Guardian. Daniel Hurst (16 July 2015), *"Bronwyn Bishop agrees to pay back cost of $5,000 helicopter trip".*

29   ABC News. 2 August 2015, *"Bronwyn Bishop resigns as Speaker, Prime Minister Tony Abbott says".*

30   *Sykes v Cleary* [1992] HCA 60, (1992) 176 CLR 77, Court of Disputed Returns (Australia)

31   Canberra Times. Andrew Brown, (10 March 2018). *"Referendum to change section 44 should be held at election, law expert says".*

32   ABC News. Stephanie Anderson, (5 January 2017). *"Centrelink debt recovery system failures have 'frightened' recipients, Andrew Wilkie says".*

33   ABC News. Henry Belot, (17 January 2017). *"Centrelink's controversial data matching program to target pensioners and disabled, Labor calls for suspension".*

34   Katharine Murphy, (17 September 2019). *"Robodebt class action: Shorten unveils 'David and Goliath' legal battle into Centrelink scheme".* The Guardian. ISSN 0261-3077.

35   The New Daily, (May 2020), *'Federal government to clear all Centrelink robo-debts, refund money'.*

36 Australian National Audit Office. (15 January 2020), *"Award of Funding under the Community Sport Infrastructure Program".*

37 The Australian. Ben Packham, (16 January 2020), *"Bridget McKenzie attacks 'ridiculous' Labor claims on $100m grants".*

38 ABC News, (2 February 2020). Bridget McKenzie quits frontbench over report she breached ministerial standards

39 https://www.aph.gov.au/Parliamentary_Business/Statistics/Senate_StatsNet/legislation/passed

40 Note: Unknown source.

41 The Scanlon Foundation, (2019), Mapping Social Cohesion

42 Professor Andrew Markus, (2019), *'Mapping Social Cohesion,'* The Scanlon Foundation Surveys

43 https://www.anu.edu.au/news/all-news/trust-in-government-hits-all-time-low (Accessed on 3/6/2020)

44 The New Daily, (June 2020), *'Australia in first recession in 29 years, Josh Frydenberg confirms.'*

45 ABS, (2019), Estimates of Industry Multifactor Productivity 2018-2019

46 World Economic Forum, (October 2019), *'Global Competitiveness Report 2019: How to end a lost decade of productivity growth'*

47 OECD Interim Economic Assessment, (2 March 2020) *'Coronavirus: The world economy at risk'*

48 Reserve Bank of Australia, (March 2019), Media Release: *'Statement by Philip Lowe, Governor: Monetary Policy Decision'*

49 Australian Government Treasury, (May 2020), *'Economic Response to the Coronavirus'*

50 Prime Minister, Treasurer, (12 March 2020), Media Release https://www.pm.gov.au/media/economic-stimulus-package

51 Roy Morgan, (16 March 2020), 'Over 60% of Australian businesses affected by COVID-19 – up from only 15% in mid-February', Article No. 8328

52 ABS, (May 2020), Media Release: *'One Third of Accommodation & food services jobs lost'*

53 Daily Mail Australia, (April 2020), *'Spring instead of next month – leaving 2.3 million people out of work'*

54 ABC News, (May 2020), *'National Cabinet to bring forward meeting on lifting coronavirus restrictions, Prime Minister Scott Morrison says'*

55 Australian Financial Review, (May 2020), *'Decade of debt sets mammoth reform task for Frydenberg'*

56 Schwab, Klaus (2017) (2016). *"The Fourth Industrial Revolution"* Crown Publishing Group

57 Rana Foroohar Australian Financial Review, (May 2020), *'Economists need to abandon their comfort zones to deal with COVID-19'*

58 OECD, (March 2020), *'Integrative Economics'*

59 The Leader, (January 2020), *'Scott Morrison responds to fire fighting chiefs' criticism of lack of communication'*

60 BBC News, (January 2020), *'Shane Fitzsimmons: 'Tireless' fire chief steering Australians through disaster'*

61 The Lowy Institute, (2019), Australia and Climate Change

62 Professor Ross Garnaut (2008) The Garnaut Climate Change Review – Final Report page 118

63 https://www.sbs.com.au/news/how-a-climate-change-study-from-12-years-ago-warned-of-this-horror-bushfire-season

64 CSIRO Submission 09/355, (July 2009) *'Bushfires in Australia'* Prepared for the 2009 Senate Inquiry into Bushfires'

65 World Economic Forum, (January 2020), *'Global Risks Report 2020'*

66 InterGovernmental Panel on Climate Change, (October 2014), *'Fifth Assessment Report'*

67 InterGovernmental Panel on Climate Change, (October 2018), *'Global Warming of 1.5' C'*

68 Richard Flanagan, (January 3, 2020), *'Australia is commiting climate suicide'*

69 https://www.abc.net.au/news/2020-01-18/australia-fires-google-search-surge-tops-decade/11874778

70 ABC News, (March 2020), *'The size of Australia's bushfire crisis captured in five big numbers'*

71 https://www.abc.net.au/news/2019-11-14/former-fire-chief-calls-out-pm-over-refusal-of-meeting/11705330

72 https://www.climatecouncil.org.au/wp-content/uploads/2019/04/fire-chiefs-statement-pages.pdf

73 https://www.dailymail.co.uk/news/article-7816335/Deputy-PM-charge-Scott-Morrison-Hawaii-slammed-blaming-bushfires-manure.html

74 ABC News, (October 2012), *'As it happened: Australia in the Asian century'*

75 Australian Government, (October 2012), *'Australia in the Asian Century – White Paper'*

76 The Lowy Institute Poll, (2019), Feelings towards other nations

77 Australian Government, Department of Foreign Affairs and Trade (https://www.dfat.gov.au/) (Accessed on 8/5/2020)

78   Australian Government, Department of Foreign Affairs and Trade (https://www.dfat.gov.au/) (Accessed on 8/5/2020)

79   Australian Government, Department of Foreign Affairs and Trade (https://www.dfat.gov.au/) (Accessed on 8/5/2020)

80   The Lowy Institute Poll, (2019), Feelings towards other nations

81   Australian Government, Department of Foreign Affairs and Trade (https://www.dfat.gov.au/) (Accessed on 8/5/2020)

82   Australian Government, Department of Foreign Affairs and Trade (https://www.dfat.gov.au/) (Accessed on 8/5/2020)

83   Cricket Australia, (http://www.cricket.com.au) Accessed on 8 May 2020

84   Australian Government, Department of Foreign Affairs and Trade (https://www.dfat.gov.au/) (Accessed on 8/5/2020)

85   Australian Government, Department of Foreign Affairs and Trade (https://www.dfat.gov.au/) (Accessed on 8/5/2020)

86   Australian Government, Department of Foreign Affairs and Trade (https://www.dfat.gov.au/) (Accessed on 8/5/2020)

87   The Lowy Institute Poll, (2019), *'Feelings towards other nations'*

88   https://www.smh.com.au/politics/federal/morrison-says-he-s-not-waiting-by-the-phone-for-invitation-to-china-20191004-p52xjd.html

89   McKinsey Global Institute, (July 2019), *'Asia's future is now'*

90   Financial Times, (March 2019), *'The Asian Century is set to begin'*

91   The Lowy Institute, (2018) *'Asia Power Index'*

92   https://www.ft.com/content/f1611686-67fe-11e9-9adc-98bf1d35a056

93   Australian Financial Review, Phillip Coorey, (29 March 2020), *'China spree sparks FIRB crackdown.'*

94   Australian Government, Department of Foreign Affairs and Trade (2017), *'2017 Foreign Policy White Paper'*

95   Worldometer – https://www.worldometers.info/world-population/ Accessed on 8/5/2020

96   Gratton, L and Scott A., *'The 100-Year Life: Living and Working in an Age of Longevity'* (Bloomsbury 2016)

97   ABS, (November 2018), *'Australia's population to reach 30 million in 11 to 15 years'* (Media Release)

98   ABS, (November 2018), *'Population Projections, Australia, 2017'* (base) – 2066 Catalogue Number 3222.0

99   ABS, (November 2018), *'Population Projections, Australia, 2017'* (base) – 2066 Catalogue Number 3222.0

100 Australian Government, *'Australian Institute of Family Studies, 'Births in Australia'*

101 Australian Government, (2020), Australian Institute of Family Studies, *'More young people living at home with parents'*

102 ABS, (April 2020), Labour Force, Australia, Mar 2020 Catalogue Number 6202.0

103 Australian Institute of Family Studies, (December 2019), *'Growing Up in Australia: The Longitudinal Study of Australian Children (LSAC) Annual Statistical Report 2018'*

104 Health Direct, (June 2019), *'Youth Suicide'*

105 Carlisle E., Fildes, J., Hall, S., Perrens, B., Perdriau, A., and Plummer, J. 2019, Youth Survey Report 2019, Sydney, NSW: Mission Australia

106 Oxford Dictionary (2017), *'Word of the Year 2017 is....'*

107 Vanity Fair, (2017), *'How a 52-Year-Old Word Invented by a Vogue Editor Became 2017's Word of the Year'*

108 McKenzie, Scott (1967) *"San Francisco (Be Sure to Wear Some Flowers in Your Hair)"* Writers: John Edmund, Andrew

109 https://www.youtube.com/watch?v=dgf0OUsQJuA&sns=em

110 ABS, (April 2020), Catalogue Number 3412.0 – Migration, Australia, 2018-19.

111 United Nations, Department of Economic and Social Affairs, Population Division (2019). 'World Population Prospects 2019'

112 United Nations, (2015), *'World Population Aging'*

113 Australia's Ageing Population, (February 2019), *'Understanding the fiscal impacts over the coming decade'* Parliamentary Budget Office

114 OECD (November 2017) *'Health at a Glance – OECD Indicators'*

115 Productivity Commission and Melbourne Institute of Applied Economic and Social Research (1999), 'Policy Implications of the Ageing of Australia's Population' Conference Proceedings, AusInfo, Canberra

116 OECD, (2013) *'Trends Shaping Education Spotlight 1 Aging Societies'*

117 Bank of America Merrill Lynch (July 2015), 'Thematic Investing Generation Next – Millennials Primer

118 AT Kearney (July 2016), *'Where are the Global Millennials'*

119 United Nations, (2015), *'World Population Aging'*

120 AFR, (June 2019), *'Finding the right formulas to drive jobs and growth'*

121 R. Botsman, (2017), *'Who Can You Trust? – How Technology Brought Us Together and Why It Might Drive Us Apart'* (Public Affairs New York)

122 Edelman, (January 2020) *'2020 Edelman Trust Barometer'*

123 Roy Morgan, (November 2019) *'Bunnings, ALDI and Woolworths on top in Net Trust Scores'* Finding No. 8199

124 Edelman Trust Barometer 2020, (March 2020), *'Special Report – Trust and the Coronavirus'*

125 R. Foster, Yale University; *'Babson Olin Graduate School of Business 2011'*

126 The Times (November 2017), *'Don't mess with Marianna Mazzucato, the world's scariest economist'*

127 Mariana Mazzucato, (2018), *'The Value of Everything'* (Penguin Books)

128 Marianna Mazzucato, (2013), *'The Entrepreneurial State. Debunking Public vs. Private Sector Myths'*

129 Australian Government, (February 2019) *'Royal commission into the Banking, Superannuation and Financial Services Industry'*

130 Australian Government, (April 2019), *'The extent and causes of the wage growth slowdown in Australia'*

131 PricewaterhouseCoopers, (2019), *'Australia Matters'*

132 PricewaterhouseCoopers (November 2019), Media release: *'Worker underpayments and digital skills deficit among top five sleeper issues for business in 2020'*

133 World Economic Forum, (January 2020), *'The Reskilling Revolution: Better Skills, Better Jobs, Better Education for a Billion People by 2030'*

134 Optus Business, (2018), *'Bridging Australia's Knowledge Economy Gap in Industry 4.0'*

135 Klaus Schwab, (2017) (2016). *"The Fourth Industrial Revolution"* (Crown Publishing Group)

136 World Economic Forum, (January 2020), *'The Reskilling Revolution: Better Skills, Better Jobs, Better Education for a Billion People by 2030'*

137 World Economic Forum, (September 2018) *'The Future of Jobs Report'*

138 Thomas Homer-Dixon, (August2002) *'The Ingenuity Gap'* (First Vintage Books Edition)

139 D. Braue, (November 2019), *'Australia needs 161,000 AI specialists by 2030'* Australian Computer Society

140 C. Tonkin, (September 2019), *'The tech talent shortage is real'* Australian Computer Society

141 Deloitte, (October 2018), *'The Industry 4.0 Paradox'*

142 OECD Better Life Initiative, (2018), *'How's Life in Australia'*

143 HRD, (June 2019), *'How widespread is workplace discrimination?'*

144 BCG (January 2018), *'The CFO's Vital Role in Corporate Transformation'*

145 McKinsey & Company (January 2018), *'Why digital strategies fail'*

146 Singularity University, Peter Diamandis (January 2020), *'Business models reshaping how we work, live and create value'*

147 Schwab, Klaus (2017) (2016). *"The Fourth Industrial Revolution"* (Crown Publishing Group)

148 Deloitte, (April 2020), *'Workforce strategies for post-COVID Recovery'*

149 L. Briscoe, (2016) *'Indigenous science is at the core of social, economic and political change'*

150 World Intellectual Property Organisation, (2019), *'Global Innovation Index 2019'*

151 EY, (March 2019), *'Agricultural Innovation – A National Approach to Grow Australia's Future'*

152 Australian Government Canberra, Innovation and Science Australia, (November 2017), *'Australia 2030: Prosperity through Innovation'*

153 Australian Government, (February 2019), *'STEM Occupations List'*

154 ABS, (March 2020), *'Labour Force, Australia'* Catalogue Number 6202.0

155 ABS, Labour Force, Australia, cat. no. 6202.0, May 2019, seasonally adjusted data.

156 Australian Government (2018), *'Industry Outlook'*

157 Australian Government, Department of Industry, Science, Energy and Resources (accessed on 18 April 2020)

158 AlphaBeta Strategy & Economics, (January 2020), *'Australian Business Investment in Innovation: levels, trends and drivers'*

159 Australian Government, (March 2020), Department of Finance *'Statistics on Australian Government Procurement Contracts'*

160 ABC, (July 2019), *'Public sector propping up employment and the economy, analysts say'*

161 ABS, (September 2019), *'Research and Experimental Development, Businesses, Australia 2017-18'* Catalogue Number 8104.0

162 BDO Australia, (May 2019), *'Federal Budget 2019-20 – BDO's Federal Election Comment'*

163 AFR, (July 2019), *'Companies clash on Australia's R&D underspend'*

164 Ray Kurzweil, (2006) *'The Singularity is Near'* (Penguin Publishing)

165 Ray Kurzweil, (1999), *'The Age of Spiritual Machines'* (Viking Press)

166 Ray Kurzweil, (March 2001), *'The Law of Accelerated Returns'*

167 World Economic Forum, (September 2015), *'Deep Shift Technology Tipping Points and Societal impact'*

168 IDC (April 2017), *'Data Age 2025: The Evolution of Data to Life-Critical'*

169 Australian Farmers Federation, (Feb 2018), NFF Code sets the ground rules for farm data' Press Release

170 IDC (April 2017), *'Data Age 2025: The Evolution of Data to Life-Critical'*

171 CISCO, (February 2018), *'CISCO Edge-to-Enterprise IoT Analytics for Electric Utilities Solution Overview'*

172 ABC News, (April 2015), *'3D printing will have a bigger economic impact than the internet, technology specialists say'*

173 McKinsey Global Institute, (February 2019), *'Mobility's second great inflection point'*

174 Digital Trends, (February 2020), *'Space X to put 42,000 satellites into orbit could face a big legal roadblock'*

175 Australian Government, (January 2017), *'Australia Day Address 2017 by Professor Michelle Y. Simmons'*

www.ingramcontent.com/pod-product-compliance
Lightning Source LLC
Chambersburg PA
CBHW061207220326
41597CB00015BA/1551